# Neuroscience Imaging Research Trends

# NEUROSCIENCE IMAGING RESEARCH TRENDS

**B. SCHALLER**
EDITOR

**Nova Science Publishers, Inc.**
*New York*

Copyright © 2008 by Nova Science Publishers, Inc.

**All rights reserved.** No part of this book may be reproduced, stored in a retrieval system or transmitted in any form or by any means: electronic, electrostatic, magnetic, tape, mechanical photocopying, recording or otherwise without the written permission of the Publisher.

For permission to use material from this book please contact us:
Telephone 631-231-7269; Fax 631-231-8175
Web Site: http://www.novapublishers.com

## NOTICE TO THE READER

The Publisher has taken reasonable care in the preparation of this book, but makes no expressed or implied warranty of any kind and assumes no responsibility for any errors or omissions. No liability is assumed for incidental or consequential damages in connection with or arising out of information contained in this book. The Publisher shall not be liable for any special, consequential, or exemplary damages resulting, in whole or in part, from the readers' use of, or reliance upon, this material.

Independent verification should be sought for any data, advice or recommendations contained in this book. In addition, no responsibility is assumed by the publisher for any injury and/or damage to persons or property arising from any methods, products, instructions, ideas or otherwise contained in this publication.

This publication is designed to provide accurate and authoritative information with regard to the subject matter covered herein. It is sold with the clear understanding that the Publisher is not engaged in rendering legal or any other professional services. If legal or any other expert assistance is required, the services of a competent person should be sought. FROM A DECLARATION OF PARTICIPANTS JOINTLY ADOPTED BY A COMMITTEE OF THE AMERICAN BAR ASSOCIATION AND A COMMITTEE OF PUBLISHERS.

**LIBRARY OF CONGRESS CATALOGING-IN-PUBLICATION DATA**

Neuroscience imaging research trends / B. Schaller, editor.
    p. ; cm.
 Includes bibliographical references and index.
 ISBN 978-1-60456-227-9 (hardcover)
 1. Brain--Imaging. 2. Neurosciences. 3. Brain--Diseases--Diagnosis. I. Schaller, Bernhard.
  [DNLM: 1. Brain--physiopathology. 2. Diagnostic Imaging--methods. 3. Brain--physiology. 4. Brain Injuries--diagnosis. 5. Diagnostic Techniques, Neurological. 6. Mental Disorders--diagnosis. WL 141 N49925 2007]
QP376.6.N48 2007
612.8--dc22
                     2007047054

*Published by Nova Science Publishers, Inc. New York*

## CONTENTS

**Preface** vii

**Editorial Articles**

**Editorial A.** Ischemic Postconditioning in the Brain.
The Breakthrough in Stroke Treatment? 1
*B. Schaller and M. Buchfelder*

**Editorial B.** Estimation of Stroke Risk in Carotid Endarterectomy:
Sense or Non-Sense of Imaging Methods? 3
*B. Schaller*

**Commentary:**

Tissue Engineering: Hype or Hope? 5
*B. Schaller and M. Buchfelder*

**Research and Review Articles**

**Chapter 1** Cerebral Representation of Hyperalgesia:
Evidence from Functional Imaging in Humans and Animals 7
*Marina Sergejeva and Andreas Hess*

**Chapter 2** Comparing the Limits of 3D Contrast Enhanced
MRA at 1.5 and 3T to Achieve High Spatial Resolution 31
*Osama Al-Kwifi and Graham A. Wright*

**Chapter 3** Rapid Estimation of Motion Blur for Real-Time
Optimization of High Resolution Imaging 41
*Osama Al-Kwifi and Graham A. Wright*

**Chapter 4** Electric Source Imaging as a Technique to Quantitatively
Determine Onset and Spread of Interictal and
Ictal Epileptiform Activity 47
*Göran Lantz, Pål G. Larsson and Christoph M. Michel*

| | | |
|---|---|---|
| **Chapter 5** | Customized Tissue Engineering in Head and Neck<br>*R. Staudenmaier, J. Welter, G. Meier and M.M. Wenzel* | 69 |
| **Chapter 6** | Recent Developments in Brain Imaging of Schizophrenia:<br>A Selective Review<br>*Vince Calhoun and Godfrey D. Pearlson* | 93 |
| **Chapter 7** | Trigemino-Cardiac Reflex and Imaging –<br>The Promise of New Insights<br>*B. Schaller, A. Filis and M. Buchfelder* | 109 |
| **Chapter 8** | The Role of Imaging in Neurological Facial Pain<br>*S. E. J. Connor* | 111 |
| **Chapter 9** | SPECT and Psychiatry – An Insight into Functional Imaging of Major Depression and Post-Traumatic Stress Disorder<br>*Marco Pagani and Ann Gardner* | 129 |
| **Chapter 10** | Comparing 1.5T and 3T Bold fMRI Imaging of Finger Tapping with Familiar and Novel Sequences<br>*Lars Nyberg, Anne Larsson, Johan Eriksson, Richard Birgander, Torbjörn Sundström and Katrin Riklund Åhlström* | 161 |
| **Index** | | 173 |

# PREFACE

This new book focuses on advances in imaging and mapping strategies to study the brain's structure, function and the relationship between both, from the whole brain to the molecular and cellular tissue level in order to improve the understanding of normal and disease processes. Studies of intact living organisms may be at the human, animal, cellular or molecular level, which advance our understanding of biological events in living systems and how these events relate to normal and pathological processes. Imaging modalities include nuclear medicine techniques (SPECT and PET) and non-nuclear techniques such as MRI, MRS, CT, ultrasound, intravital microscopy, optical imaging, diffuse optical tomography, electromagnetic tomography and other methods which elucidate molecular and cellular mechanisms, accelerate the understanding of biology, test the efficacy of therapeutic interventions in intact living systems and assess the therapeutic outcomes.

Chapter 1 - *Background:* Functional imaging of the brain, as the highest station of pain processing, is already acknowledged as an efficient tool in pain research. Many human studies also addressed hyperalgesia. On the contrary, there are only a few analogical animal studies until now, although small animal imaging (SAI) could be very helpful for investigation of this clinically relevant pain syndrome.

*Objective:* To systematically review human and rat studies regarding functional brain imaging during different forms of hyperalgesia, trying to distinguish brain activation changes caused by hyperalgesia *per se* from those caused by normal pain, and estimate the validity of SAI for studying hyperalgesia.

*Search Strategy:* PubMed searches were performed to find all articles with English-language abstracts concerning brain activity during hyperalgesia revealed by functional imaging methods - fMRI, PET, SPECT, EEG, MEG, MRS, MSI, autoradiography - till the end of the year 2006. Additional studies were found from bibliographies.

*Study Selection:* From the identified articles excluded were: those where the detected brain activation was not ascribed to any definite brain structures, studies regarding only visceral hyperalgesia, studies at purely molecular level, autoradiography studies with no additional correlation analysis between behavioural hyperalgesia signs and metabolic changes.

*Study Description and meta-analysis:* The included studies were classified according to the hyperalgesia form investigated: heat (10 human and 3 rat studies), light brush (11 and 1), pin-prick (6 and 2), light brush+pin-prick (1 and 0), cold rubbing (4 and 0), pressure (4 and 0), chemical (0 and 1) and "virtual" (1 and 0). Common and differential features of brain activation patterns during these hyperalgesia forms are revealed and discussed. A comparison between rat and human data was made within categories.

*Conclusions:* A prominent involvement of affective-motivational, cognitive and limbic circuits appears as a distinctive feature of hyperalgesic pain processing. This is especially conspicuous by hyperalgesia devoid of any mechanical component. The findings of the rat hyperalgesia studies supported those of the human ones, thus validating SAI as a useful tool for investigation of this clinical pain syndrome.

Chapter 2 - *Purpose:* To investigate signal-to-noise ratio (SNR) behavior at 1.5T and 3T using different spatial resolutions with 3D contrast-enhanced magnetic resonance angiography (CE MRA) and to evaluate the role of an 8-channel coil and the parallel imaging technique in trading off SNR and scan time.

*Materials and Methods:* An automatically triggered 3D CE MRA protocol was performed at both field strengths. The intracranial arteries were imaged under four different scenarios, changing the spatial resolution and coil setting.

*Results:* SNR is doubled at 3T compared with 1.5T. At higher spatial resolution (0.5mm isotropic) SNR drops dramatically at both field strengths, as expected. Image quality at 1.5T becomes unacceptable for diagnosis, whereas images at 3T are acceptable. The 8-channel coil is limited in improving SNR values in deep intracranial arteries at 1.5T.

*Conclusion:* Obtaining a 0.5mm isotropic spatial resolution is feasible at 3T, as a result of higher SNR. Parallel imaging may facilitate scan time reduction at 3T for high resolution acquisitions.

Chapter 3 - Due to substantial carotid and coronary motion variation among individuals it is essential that optimal imaging parameters be determined by characterizing this motion on a subject-by-subject basis. In this work, new methods to compute the effects of motion on different k-space trajectories are introduced. After measuring in-vivo motion a simulation was performed to calculate motion blurring effects using different methods. Final results from each method were evaluated for their accuracy and speed in estimating the motion effects. It has been demonstrated that, for small motions corresponding to reasonable target resolutions, simple approximation methods yield reasonable estimates of motion blur significantly faster than full computations of the point spread function, and could be used to facilitate real-time scan parameter optimization to achieve high resolution imaging.

Chapter 4 - In this paper the recent development in Electric Source Imaging (ESI) in epilepsy investigations will be discussed. ESI has made tremendous progress in the past years and is now available as an additional tool in the presurgical workup of patients with pharmaco resistant partial epilepsy. Many different source models have been proposed, each with their advantages and disadvantages, and it is therefore advisable to perform the source reconstructions not with one single model but rather with a number of these different techniques. It is also important that the reconstructions are performed using a realistic head model, preferably the MRI of the individual patient, and that a sufficient number of recording electrodes (at least 60-70) are used. The results can also be improved by using statistical SPM like techniques to determine areas of significant activation. In the paper these different methodological issues are addressed. Some of the literature on interictal and ictal source propagation is reviewed, and it is discussed how ESI can be used to characterize the focus of interictal and ictal epileptiform activity, in terms of focus localization but also in terms of visualizing these complex propagation patterns. Accurately determining the anatomincal origin of interictal and ictal epileptiform activity from EEG recordings is of crucial importance in presurgical epilepsy investigations, and the possibility to characterize interictal and ictal propagation patterns also gives important additional information for understanding the epileptic process in the individual patient, for instance in relation to the ictal semiology or to the results of other neuroimaging techniques.

Chapter 5 - Tissue Engineering (TE) of cartilage for reconstructive surgery seems to be a promising option to gain tissue for three-dimensional (3D) structures with a minimal donor site morbidity. A specific defect needs a customized implant. Most TE strategies rely on the application of resorbable three-dimensional (3D) scaffolds to guide the growing tissue. Each tissue requires a specific scaffold with precisely defined properties which depend on physiological environment.

Rapid Prototyping (RP) technologies allow to fabricate scaffolds with any geometric complexity, from multiple materials, as composites and even the inner architecture of the object can be varied in a defined manner and at a defined spot. Scaffolds can be manufactured by RP techniques directly from computer added design (CAD) data sources e.g. via STL file.

The combination of Tissue Engineering and Rapid Prototyping presents the basis for the realization of customized implants. It provides a helpful support for conventional ear reconstruction and gives new perspectives for the reconstructive surgery.

Chapter 6 - The use of noninvasive brain imaging in schizophrenia has made significant advances in recent years. One key area which has grown significantly is the emphasis on the identification of image-based biological markers–which have the potential to impact the way that schizophrenia is diagnosed and treated. This is enhanced by the development of new image analysis techniques for identifying relevant imaging information and combining the information gleaned from the various imaging modalities. A third key factor here is the advent of multi-site imaging studies. Finally, a number of recent studies have examined the relationship between brain imaging and genetic data. Given the known importance of both genetics and environment in brain function, the integration of genetics with brain imaging has the potential to fundamentally change our understanding of human brain function in disease through the identification of imaging endophenotypes. In this review the authors discuss the current impact of these new areas and their future implications.

Chapter 7 - The trigemino-cardiac reflex (TCR) is a well-known entity that has gained more and more interest. The reproducible hypotension and bradycardia upon stimulation of the trigeminal nerve, has been reported during craniofacial surgery and during surgery within the cerebellopontine angle, petrosal sinus, orbit, trigeminal ganglion and the falx. Awareness of TCR allows for early detection and appropriate treatment.

Chapter 8 - This review examines the application of imaging to the neurological aspects of facial pain. Neuralgic facial pain focusing on trigeminal neuralgia, facial pain with associated cranial nerve deficits, pain referred through shared roots of cranial nerve innervation, and other neurological facial pain syndromes will be discussed. The diagnostic accuracy, diagnostic yield of imaging and the impact on clinical management in these clinical settings will be reviewed and the structural correlates of facial pain will be illustrated.

Chapter 9 - During the last 30 years functional brain imaging has been increasingly applied to psychiatric disorders. In the field of neurodegenerative disorders, single photon emission computed tomograpghy (SPECT) and positron emission tomography (PET) allow nowadays for the identification of mild-to-severe forms of Alzheimer's Diseases (AD), Frontal Lobe Dementia (FLD) and Parkinson Disease (PD) with a sensitivity and specificity approaching 80-90%. Such high accuracy has not been realized for other forms of dementia (i.e. vascular dementias and Dementia with Lewy's Bodies), for mild cognitive impairment, for schizophrenia, for post-traumatic stress disorder or for all forms of depression in which the links between the findings of functional brain imaging studies and the neural substrates of such disorders have not been clearly established yet.

In this scenario, the importance of the combination of newly developed imaging and statistical techniques in the assessment of psychiatric disorders steadily increases. The neuropsychiatric and behavioural abnormalities are often a source of considerable patient and caregiver distress, whilst proper diagnosis also contributes to the decision of the level of care of these patients resulting in early diagnosis and better patient management with considerable savings for the community.

Optimised nuclear medicine techniques and algorithms for functional brain imaging could now be implemented in the clinical management of psychiatric patients for finer discrimination of cerebral blood flow (CBF) or metabolic changes. Such changes have for long time been neglected due to the fact that the quality of both functional images and image analysis was not sufficient to identify the sometimes small functional regional variations occurring in psychiatric diseases.

In this respect it is of utmost importance to highlight that in recent decades a general consensus on the CBF distribution pattern in AD and FLD has been built and that clinical diagnosis is now routinely based on the visual identification of reduced radiopharmaceutical uptake in the temporo-parietal cortex. Such consensus is still lacking for psychiatric disorders in which a specific CBF distribution pattern has not yet been identified.

The present review concentrates on state-of-the-art technologies and on recent investigations dealing with SPECT and psychiatry. As for neurodegenerative diseases (i.e. AD, FLD and PD) extensive reviews are widely available [1, 2, 3] and schizophrenia has recently been reviewed in this journal [4]. The authors will introduce and briefly comment on the most recent methodological advancements and statistical techniques. They will also discuss two psychiatric disorders in which functional neuroimaging has recently played a substantial role in identifying the neurobiological changes and, in perspective, in improving the clinical diagnosis.

Chapter 10 - It has been suggested that fMRI at 3T yields stronger and more extensive BOLD activations than fMRI at 1.5T, and that imaging at higher field strengths can reveal unique activations. In the present study the authors compared, within-subjects, activation patterns during a finger-tapping task at 1.5 and 3T. The data were analyzed with a random-effects model in SPM2. At a strict statistical level ($p<0.05$, FWE correction for multiple comparisons), ipsilateral cerebellar activation was revealed at 1.5T. At 3T, activation in sensory-motor regions in the contra-lateral cerebrum was identified in addition to the activation in cerebellum. At a less stringent statistical threshold, imaging at 1.5T and 3T revealed overlapping cortical regions with more extensive clusters at 3T. A similar pattern was seen in a comparison of familiar and novel sequences. However, subcortical activations of thalamus and parts of the basal ganglia were uniquely identified at 3T. Analyses at the individual level substantiated the group results by showing that the higher sensitivity of the 3T resulted in images with higher between-individual consistency in activation patterns.

*Editorial A*

# ISCHEMIC POSTCONDITIONING IN THE BRAIN. THE BREAKTHROUGH IN STROKE TREATMENT?

## B. Schaller and M. Buchfelder
### Zurich, Switzerland and Erlangen, Germany.

Ischemic preconditioning has emerged as a powerful experimental method of ameliorating ischemic-reperfusion injury in a variety of organs. Clinical trials using ischemic preconditioning have been successfully carried out . They support the existence of ischemic preconditioning induced cyto-protection in humans as well. More studies with greater patient numbers need to be carried out in these areas to demonstrate the efficacy of ischemic preconditioning in providing clinical benefit in terms of reducing morbidity and mortality in stroke treatment. Although laboratory and experimental evidence is favorable, clinical studies using ischemic preconditioning in neurovascular surgery are currently lacking.

Specifically, preconditioned tissues exhibit reduced energy requirements, altered energy metabolism, better electrolyte homeostasis and genetic re-organisation, giving rise to the concept of „ischaemia tolerance". Ischemic preconditioning also induces „reperfusion tolerance" with less reactive oxygen species and activated neutrophils released, reduced apoptosis and better microcirculatory perfusion compared to non-preconditioned tissue. Systemic ischaemia/reperfusion injury is also diminished by preconditioning. Ischemic preconditioning is ubiquitous but more research is required to fully translate these findings to the clinical arena.

Reperfusion damage is a complex process involving several cell types, soluble proinflammatory mediators, oxidants, ionic and metabolic dyshomeostasis, and cellular and molecular signals [1,2]. Novel neuroprotective strategies are required to target this form of injury [3]. The neuroprotective potential of ischemic preconditioning has not been realized in clinical practice because it necessitates an intervention applied before the onset of ischemic stroke, which is difficult to predict. A more-amenable approach to neuroprotection is to intervene at the onset of reperfusion, the timing of which is under the control of the operator. In this regard, these new findings of postconditioning in the brain may open a window to improve stroke treatment or prevention [4]. In contrast to preconditioning, which requires a foreknowledge of the ischemic event, postconditioning can be applied at the onset of reperfusion at the point of clinical service. Interestingly, experimental studies suggest that ischemic preconditioning and postconditioning activate the same signaling pathway at the time of reperfusion, thereby offering a common target for neuroprotection [2]. Therefore, the pharmacologic recruitment of this signaling pathway at the time of cerebral reperfusion might allow one to harness the neuroprotective potential of ischemic preconditioning and postconditioning and therefore substantially improve the outcome in ec-ic bypass and neurovascular surgery.

Serial brain images are necessary to visualize and document significant preservation of regional CBF in the periinfarct area, which is relevant to find candidates for the long-term inhibiting effect on infarct evolution in the tolerant state. But, we are far away from understanding the processes in this phase of stroke [5]. Clinical stroke research has become dependent on imaging in an effort to sharpen the understanding of the dynamic processes involved. It is interesting that despite the sensitivity of current imaging methods, it is not possible to identify an appropriate lesion in all patients with acute stroke deficits. But such identification is more important if postischemic therapeutic methods should be usefull in the future (4). Optimal stroke care requires collaboration between the clinician and radiologist, for we are clearly far from total understanding of the process. As improvements in technology allow further investigation, the more information we will have to be considered and applied to patient care.

Further research is needed to find the imaging correlations in humans to the best known cellular and subcellular data found in experimental animal models. Such findings would open the window to new pharmacological agents that would mimic postconditioning in order to treat all patients with ongoing acute ischemic stroke. Of course, the potential of postconditioning must be rigorously tested in clinical trials, first for its safety and feasibility and then subsequently for its efficacy and therapeutic potential. It will be mandatory in such clinical trials to carefully control for confounding variables such as the size of the area at risk, the duration of the preceding ischemic insult, and collateral status. Neglect of these confounding variables has probably contributed to the failure of translation of experimentally validated principles of neuroprotection to the clinical arena (e.g., adenosine receptor activation) in the past

## REFERENCES

[1]  Schaller B, and Buchfelder M. Editorial. The janus shape of glucose and brain under stress conditions. *Neuroscience Imaging.* 2006; 1:229-230.

[2]  Zhao H, Sapolsky RM, Steinberg GK. Interrupting reperfusion as a stroke therapy: ischemic postconditioning reduces infarct size after focal ischemia in rats. *J. Cereb. Blood Flow Metab.* 2006;26:1114-21.

[3]  Schaller B. The role of endothelin in stroke: Experimental data and underlying pathophysiology. *Arch. Med. Sci.* 2006; 2:146-158

[4]  Schaller B, Graf R, Jacobs AH. Ischaemic tolerance a window to endogenous neuroprotection? *Lancet.* 2003; 362:1007-8.

[5]  Gress DR. Imaging in stroke: the more you look, the more you see. *AJNR Am. J. Neuroradiol.* 2000;21:1569.

*Editorial B*

# ESTIMATION OF STROKE RISK IN CAROTID ENDARTERECTOMY: SENSE OR NON-SENSE OF IMAGING METHODS?

## *B. Schaller*

Carotid endarterectomy (CEA) is performed to prevent embolic stroke in patients with atheromatous disease at the carotid bifurcation. There is now substantial evidence to support early operation in symptomatic patients, ideally within 2 weeks of the last neurological symptoms. Thus, the anaesthetist may be faced with a high risk patient in whom there has been limited time for preoperative preparation. Recently, Press et al. [1] described a definite benefit from carotid endarterectomy-specific risk model predicting a broad range of medical, neurological, and surgical complications following surgery for carotid artery stenosis.

Estimation of stroke risk after surgery for carotid artery stenosis is similar whether statewide data or institution-specific data were used [2]. The state-wide model is applicable to institution-specific data collected over several years. As demonstrated by the authors as well as by others, modelling stroke risk after surgery for carotid artery stenosis is possible. Such models may be used to identify patient subgroups at increased risk for stroke after surgery for carotid artery stenosis. However, two operative techniques (use of local anaesthesia and patch closure) are best-known to lower the risk of death or stroke. The authors caution that overall prognostic performances of such risk models are modest and that the balance between risk and benefit might differ in daily practice because high mortality during the procedure or rigorous medical treatment could diminish the beneficial effects of intervention in asymptomatic patients. This dilemma is further exaggerated when a decision is to be made in patients with stenoses less than 70%. Therefore, there is a need to differentiate high risk, unstable atherosclerotic plaque biology from stable atherosclerosis [3].

It is a matter of debate whether a multimodal assessment of plaque vulnerability involving the combination of systemic surrogate markers, new imaging methods that target inflammatory and thrombotic components, and the potential of emerging therapies may lead to a new stratification system for atherothrombotic risk and to a better prevention of atherothrombotic stroke after surgery for carotid artery stenosis in the near future [4]. Such a line of action may not only help improve individual patient outcomes, they also may help direct scarce medical resources to maximize medical benefits, improve overall medical care, and minimize costs and untoward side effects. The cardiac risk models that are investigated by the authors [1] are only one but an important part of such a concept.

To our current knowledge, the imaging methods play an important role in risk for postoperative ischemic events after carotid endarterectomy. The hemodynamic effects of carotid endarterectomy on the collateral blood supply and on the regional cerebral blood flow (rCBF) have not been established. Recently, arterial spin-labeling magnetic resonance imaging has been introduced as the first method to quantify the actual territorial contribution of individual collateral arteries as well as to noninvasively measure rCBF. However, collateral circulation plays a major role in maintaining CBF in patients with internal carotid artery stenosis. CBF can remain normal despite severe internal carotid artery stenosis, making the benefit of carotid endarterectomy difficult to assess. The unilateral arterial spin labeling technique provides a method for isolating the CBF supply from the left and right side of the neck, providing information about cerebral hemodynamics not available from conventional bilateral arterial spin labeling methods such as direct measurement of cerebral hypoperfusion resulting from internal carotid artery stenosis, the amount of collateral supply to the hemisphere on the side of a stenosis, and the redistribution of collateral supply post- carotid endarterectomy. The best predictor of increased CBF in the MCA territory on the side of surgery seems to be the preoperative CBF in the same territory. These new thechnique demonstrate in an impressive manner how small our knowledge is about cerebral hemodynamic. Lets go about the research on it!

## REFERENCES

[1] Press MJ, Chassin MR, Wang J, et al. Predicting medical and surgical complications of carotid endarterectomy. Comparing the risk indexes. *Arch. Intern. Med.* 2006;166:914-920.

[2] Ricotta JJ, Char DJ, Cuadra SA, et al. Modeling stroke risk after coronary artery bypass and combined coronary artery bypass and carotid endarterectomy. *Stroke.* 2003; 34:1212.

[3] Kietselaer BL, Hofstra L, Narula J. ACST: which subgroups will benefit most from carotid endarterectomy? *Lancet.* 2004;364:1124-5.

[4] Nighoghossian N, Derex L, Douek P. The vulnerable carotid artery plaque: current imaging methods and new perspectives. *Stroke.* 2005;36:2764-72.

[5] Schaller B, Buchfelder M: Ischemic Postconditioning in the Brain: The Breakthrough in Stroke Treatment? *Neuroscience Imaging* 2006: 1: 231-232

[6] Schaller B, Buchfelder M. The Janus Shape of Glucose and Brain Under Stress Conditions. *Neuroscience Imaging* 2006; 1: 229-230.

[7] Jongen C. Analyzing Group Differences in Stroke Pattern. *Neuroscience Imaging* 2006; 1: 211-213.

*Commentary*

# TISSUE ENGINEERING: HYPE OR HOPE?

## *B. Schaller and M. Buchfelder*
Zurich, Switzerland and Erlangen, Germany.

The adoption of combinatorial and computational methods in biomaterials design is a highway towards the discovery and realization of tailored polymeric materials to satisfy the specific requirements of many diverse biomedical or prosthetic applications. These laboratory techniques may be useful in developing therapies for replacing lost tissue function, as in vitro models of living tissue, and also for further enabling fundamental studies of structure/function relationships in three dimensional contexts [1].

Staudenmaier et al. [2] introduce their concept that allows among others total ear reconstruction in two surgical steps with special reference to the underlying imaging methods. The paucity of techniques and materials emphasizes the need for alternative bone formation strategies in state-of-the-art medicine. Recent integrative approaches suggest that successful reconstruction requires interdisciplinary teams, with surgeons interacting with imaging experts, materials scientists, and engineers. These new techniques are also of special interest for the imaging specialist, not only because of the need of such specialist in the interdisciplinary teams, but also because of the facts that this research may open a window to better understand the behavior of different tissues.

Living cells can sense mechanical forces and convert them into different biological responses. Similarly, biological and biochemical signals are known to influence the abilities of cells to sense, generate and bear mechanical forces. Studies into the mechanics of single cells, subcellular components and biological molecules have rapidly evolved during the past decade with significant implications for biotechnology and human health. This progress has been facilitated by new capabilities for measuring forces and displacements with piconewton and nanometre resolutions, respectively, and by improvements in neuroscience imaging. Details of mechanical, chemical and biological interactions in cells remain elusive. However, the mechanical deformation of proteins and nucleic acids may provide key insights for understanding the changes in cellular structure, response and function under force, and offer new opportunities for the diagnosis and treatment of disease. Such research focused on deciphering the biochemical mechanisms that regulate cell proliferation and function has largely depended on the use of tissue culture methods in which cells are grown on two-dimensional (2D) plastic or glass surfaces. However, the flat surface of the tissue culture plate represents a poor topological approximation of the more complex three-dimensional (3D) architecture of the extracellular matrix (ECM) and the basement membrane (BM), a structurally compact form of the ECM. Recent work has provided strong evidence that the highly porous nanotopography that results from the 3D associations of ECM and BM nanofibrils is essential for the reproduction of physiological patterns of cell adherence, cytoskeletal organization, migration, signal transduction, morphogenesis, and differentiation in cell culture. In vitro approximations of these nanostructured surfaces are therefore desirable for more physiologically mimetic model systems to study both normal and abnormal functions of cells, tissues, and organs. In addition, the development of 3D culture environments is imperative to achieve more accurate cell-based assays of drug sensitivity, high-throughput drug discovery assays, and in vivo and ex vivo growth of tissues for applications in regenerative medicine.

In summary, the learned principles of regenerative medicine are beginning to be applied successfully and are a step closer to human clinical applications. We now understand that if the appropriate mixture of cells, substrates, and environment are provided, then the in vivo self-assembly of functional tissue structures is possible. The result of such a process in the present report of Staudenmaier et al. [2] is not the perfect recapitulation of development of a normal tissue in head and neck surgery but rather an acceptable biological substitute that can become integrated into host tissues and will likely continue to remodel with time.

## REFERENCES

[1] Illes J, Atlas, SW and Raffin TA. Imaging neuroethics for the imaging neurosciences. *Neuroscience Imaging.* 2005; 1: 4.
[2] Staudenmaier R., Welter J, Meier G, and et al. Customized tissue engineering in head and neck. *Neuroscience Imaging.* 2006; 1: in press

*Chapter 1*

# CEREBRAL REPRESENTATION OF HYPERALGESIA: EVIDENCE FROM FUNCTIONAL IMAGING IN HUMANS AND ANIMALS

*Marina Sergejeva and Andreas Hess*[*]

Department of Experimental and Clinical Pharmacology and Toxicology,
Fahrstrasse 17, D-91054, Erlangen, Germany

## ABSTRACT

*Background:* Functional imaging of the brain, as the highest station of pain processing, is already acknowledged as an efficient tool in pain research. Many human studies also addressed hyperalgesia. On the contrary, there are only a few analogical animal studies until now, although small animal imaging (SAI) could be very helpful for investigation of this clinically relevant pain syndrome.

*Objective:* To systematically review human and rat studies regarding functional brain imaging during different forms of hyperalgesia, trying to distinguish brain activation changes caused by hyperalgesia *per se* from those caused by normal pain, and estimate the validity of SAI for studying hyperalgesia.

*Search Strategy:* PubMed searches were performed to find all articles with English-language abstracts concerning brain activity during hyperalgesia revealed by functional imaging methods - fMRI, PET, SPECT, EEG, MEG, MRS, MSI, autoradiography - till the end of the year 2006. Additional studies were found from bibliographies.

*Study Selection:* From the identified articles excluded were: those where the detected brain activation was not ascribed to any definite brain structures, studies regarding only visceral hyperalgesia, studies at purely molecular level, autoradiography studies with no additional correlation analysis between behavioural hyperalgesia signs and metabolic changes.

*Study Description and meta-analysis:* The included studies were classified according to the hyperalgesia form investigated: heat (10 human and 3 rat studies), light brush (11 and 1), pin-prick (6 and 2), light brush+pin-prick (1 and 0), cold rubbing (4 and 0),

---

[*] Corresponding author. Tel.: +49 9131 85 22003/22801; Fax: +49 9131 85 22774; E-mail: andreas.hess@pharmakologie.uni-erlangen.de

pressure (4 and 0), chemical (0 and 1) and "virtual" (1 and 0). Common and differential features of brain activation patterns during these hyperalgesia forms are revealed and discussed. A comparison between rat and human data was made within categories.

*Conclusions:* A prominent involvement of affective-motivational, cognitive and limbic circuits appears as a distinctive feature of hyperalgesic pain processing. This is especially conspicuous by hyperalgesia devoid of any mechanical component. The findings of the rat hyperalgesia studies supported those of the human ones, thus validating SAI as a useful tool for investigation of this clinical pain syndrome.

**Keywords:** hyperalgesia, allodynia, brain, functional imaging.

# BACKGROUND

*Pain* is defined as "an unpleasant sensory and emotional experience associated with actual or potential tissue damage, or described in terms of such damage" (IASP - International Association for the Study of Pain). It is transmitted within a network of certain neuronal structures (pain matrix: Figure 1). Acute pain acts as an alarm signal to prevent tissue damage and was optimized for this purpose during the evolution. Lasting pain leads to processes of neuronal sensitization which as a first step manifests itself as *hyperalgesia* or *allodynia*. Sustained sensitization may lead to pain chronification and, as a consequence, to *abnormal pain processing*. Chronic pain has no biological advantage but may significantly reduce the quality of live. Thus, investigation of these phenomena is of a highest clinical importance.

*Functional in-vivo imaging* of the pain network has already gained an important role in basic and clinical research. A lot of knowledge about pain processing at different central nervous system (CNS) levels has been accumulated with help of functional Magnetic Resonance Imaging (fMRI), Positron Emission Tomography (PET), Magnetic Resonance Spectroscopy (MRS), Single Photon Emission Tomography (SPECT), Electroencephalography (EEG), Magnetoencephalography (MEG), Magnetic Source Imaging (MSI) (for reviews see: [1], [2], [3], [4], [5], [6], [7], [8], [9], [10], [11], [12], [13]). The majority of pain imaging studies has been performed on humans and focused on pain-induced activity of the brain, as the highest station of pain processing. Among these different methods fMRI is distinguished by its 3D-resolution, non-invasiveness, and harmfulness. Therefore, it is especially well applicable in drug evaluation, since it allows repetitive or long-lasting investigations on healthy subjects and clinical patients, as well as on animals. Although there are yet no published examples of fMRI applied to novel compounds, many studies showed quantifiable effects of existing, licensed drugs, demonstrating how fMRI can be used in future drug discovery (for reviews see: [14], [15]). Human studies aimed at imaging clinically relevant phenomena like hyperalgesia or allodynia are also well represented in the current literature (Table 1A).

On the contrary, *in-vivo* functional brain imaging studies on animal pain models are until now far not so numerous, as human studies or animal *ex-vivo* imaging autoradiographic (ARG) studies. Animal *in-vivo* imaging in general, and pain imaging especially, is complicated by the fact that in most cases experimental animals have to be anesthetized, thus confounding the basic physiology and consequently brain activity. Brain imaging studies of painful conditions in anesthetized rats illustrating modulatory influence of drugs (morphine)

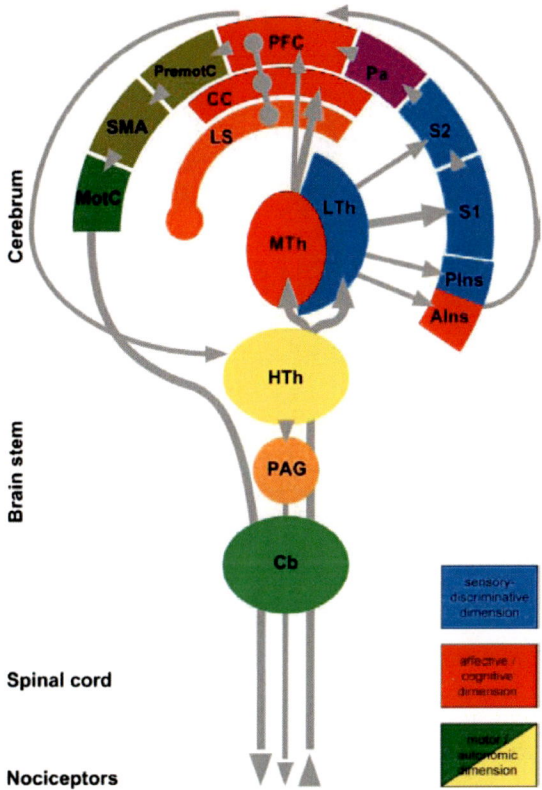

Figure 1. Schematic depiction of the nociceptive pathway. *Explanation.* Acute pain is a multidimensional sensation initiated at pain receptors in the periphery (nociceptors) by (potentially) harmful stimuli. The pain system consists of ascending and descending pathways which are highly interconnected at different processing stages up to the prefrontal cortex (PFC), as the highest station of pain processing. The most important transfer (and "preprocessing") stations of nociceptive information are: spinal cord (dorsal horn neurons), brainstem including among others medulla oblongata and midbrain peri-aqueductal grey (PAG), and thalamus. From there and upwards one differentiates between two although functionally overlapping but basically different subsystems. The *lateral thalamus (LTh)*, particularly ventral posterolateral / ventral posteromedial nuclei, projects to *somatosensory* cortex: areas *S1* and *S2*. These all constitute the so called *lateral pain system* responsible for the *sensory-discriminative* dimension of pain. Further processing of somatosensory information occurs within the *parietal cortex (Pa)*. The *medial thalamus (MTh)*, composed of intralaminar (centromedian, centrolateral and parafascicular) and midline (mediodorsal and laterodorsal) nuclei, has tight connections to the *anterior* cingulate *cortex (CC)* and further to the *PFC*. This is the *medial pain system* considered to be "in charge" of the *affective-motivational* dimension of pain. Moreover, this system has also extensive interconnections with the *limbic system (LS)*: entorhinal cortex, hippocampus, amygdala, hypothalamus, - which is inseparably associated with emotions. The *insula* has an intermediate position, since it receives somatosensory input *(posterior part, PIns )*, but has strong reciprocal connections to the amygdala *(anterior part, AIns )*. Therefore, insula can be ascribed to the medial pain system. As an "output" of pain processing, structures for immediate motor and autonomic responses and pain control are activated. The higher order motor responses originate in *PFC, premotor (PremotC) and supplementary motor (SMA)* cortical areas and next in *primary motor cortex (MotC)*. They descend downstream commands to the lower motoneurons in the spinal cord. Motor nuclei of the thalamus (ventral anterior / ventrolateral), basal ganglia (caudate putamen, nucleus accumbens) and *cerebellum (Cb)* take active part in coordination of these motor responses. *Hypothalamus (HTh)* and other components of the limbic system initiate the cascades of autonomic (vegetative) responses to pain. One of the most important parts of the descending inhibitory pain control system acts via the *PAG* upon the dorsal horn neurons.

are sparse [16], [17], [18], [19]. Nevertheless, rat brain structures, shown to be activated and modulated in these studies, generally corresponded to those reported for humans. Even less studies directly addressed clinically relevant syndromes like hyperalgesia/allodynia in animals. (cf. Table 1B). Malisza and colleagues [20] and later Moylan Governo and colleagues [21] have explored with fMRI the model of capsaicin-induced allodynia on rats.

In the recent work from our group [22] we have investigated hyperalgesia in the rat zymosan model [23].

The spectrum of well established animal pain, disease and transgenic models is very broad, including those of e.g. inflammatory and neuropathic pain, that closely mimic clinical pain states (for review see: [24]). "Visualizing" CNS events underlying these models with functional small animal imaging (SAI) should be very helpful for investigating their physiological mechanisms, testing new drugs, and forthcoming treatments.

**Table 1A.** The studies under review ordered according to imaging modality, hyperalgesia / allodynia trigger and test stimulation type: A – human studies, B – rat studies

| Imaging modality | Hyperalgesia/allodynia trigger | Test stimulus | N° of studies |
|---|---|---|---|
| | *experimental* | | |
| fMRI | capsaicin | heat | 3 |
| | anxiety | | 1 |
| | capsaicin | light brush | 2 |
| | capsaicin | pin-prick | 3 |
| | capsaicin | l.brush+p.-prick | 1 |
| PET | capsaicin | heat | 1 |
| | capsaicin | light brush | 2 |
| | hypertonic saline i.m. | pin-prick | 1 |
| MSI | capsaicin | electric stim. | 1 |
| | *clinical* | | |
| fMRI | irritable bowel syndrome | heat | 1 |
| | sympathetically mediated pain | | 1 |
| | burning mouth disorder | | 1 |
| | NS lesions | cold rubbing | 2 |
| | CRPS | light brush | 1 |
| | NS lesions | | 2 |
| | CRPS | pin-prick | 1 |
| | CLBP or FM | pressure | 4 |
| | CRPS or spinal cord injury | „virtual brush" | 1 |
| PET | rheumatoid arthritis | heat | 1 |
| | atypical facial pain | | 1 |
| | NS lesions | cold rubbing | 2 |
| | peripheral nerve injury | light brush | 2 |
| LEPs | migrane | heat | 1 |
| MEG | femoral nerve neuralgia | light brush | 1 |
| MSI | CRPS | pin-prick | 1 |

## Table 1B

| Imaging modality | Hyperalgesia/allodynia trigger | Test stimulus | N° of studies |
|---|---|---|---|
| fMRI | *experimental* zymosan | heat | 2 |
| | capsaicin i.j. | light brush | 1 |
| | capsaicin | pin-prick | 1 |
| 2DG ARG | -endorhpin system blockade | formalin | 1 |
| rCBF ARG | sciatic nerve CCI | heat | 1 |
| | | pin-prick | 1 |

## OBJECTIVE

The aims of the present study are:

a) to systematically review the literature regarding functional brain imaging during hyperalgesia, allodynia, and related to them abnormal pain processing, trying to distinguish brain activation changes caused by hyperalgesia/allodynia *per se* from those caused by normal pain;
b) to define common and/or differential aspects of these changes by different hyperalgesia/allodynia forms;
c) to compare the findings from human imaging with those from SAI and thus estimate the validity of the latter for studying clinical pain syndromes.

## METHODS

### Search Strategy and Selection Criteria

Initially PubMed searches were performed to find all articles with English-language abstracts concerning brain activity during hyperalgesia revealed by functional imaging till the end of 2006. The search combinations were "hyperalgesia" or "allodynia" and "brain" and one of the following terms: "imaging", "fMRI", "PET", "SPECT", "EEG", "MEG", "MRS", "MSI", "autoradiography". Additional related papers were found from bibliographies. From all these articles we excluded:

- those where the detected brain activation was not ascribed to any certain brain structures;
- purely anatomical studies;
- studies regarding only visceral hyperalgesia/allodynia;
- studies at purely molecular level;

- autoradiography studies with no additional correlation analysis between behavioural hyperalgesia/allodynia signs and metabolic changes (for explanation see: *Biases*).

37 human and 7 rat studies met our criteria. Note that some articles contained more than one study, for example, thermal and mechanical hyperalgesia were investigated; such cases are considered here as separate studies.

## Analysis

First, to analyze the impact of different imaging modalities on the results of the studies under review, we calculated the incidence of the brain structures, generally activated by pain, showing increased activation under hyperalgesia/allodynia conditions (cf. Figure 2). Some studies have been excluded, for instance, [21], reporting only activity decreases, and MSI, MEG and LEPs studies, because of their small numbers (2, 1 and 1, respectively).

Then, we subdivided all the studies under review into categories according to the form of hyperalgesia/allodynia investigated (see Table 2), and compared human versus rat studies within each category. The overall result of our descriptive meta-analysis is summarized in Table 3.

**Table 2. Forms of hyperalgesia occuring in the studies under review, simplified from [26].** *Afferents mediating hyperalgesia:* **A-β-LTM – Aβ-fiber low threshold mechanoreceptor (touch receptor), type I AMH – A-fiber nociceptor with slow, high-threshold heat response, probably equivalent to A-fiber high-threshold mechanoreceptor; type II AMH – A-fiber nociceptor with rapid, low-threshold heat response; CMH – C-fiber mechano-heat nociceptor; MIA – mechanically insensitive (silent) nociceptor.** *Sensitization level:* **P – peripheral, C – central**

| | primary hyperalgesia | | | secondary hyperalgesia | | |
|---|---|---|---|---|---|---|
| | afferents | sensitization | | | afferents | sensitization |
| heat | I&II AMH, CMH | P | | | | |
| light brush | Aβ-LTM | C | | light brush | Aβ-LTM | C |
| pin-prick | I AMH, MIA | P&C ? | | pin-prick | I AMH | C |
| pressure | MIA, (I AMH ?) | P | | | | |
| chemical | II AMH, CMH, MIA ? | P ? | | | | |
| | | | | cold ? | | C ? |

## RESULTS

### Definitions: Hyperalgesia / Allodynia

According to IASP taxonomy [25], *hyperalgesia* means an enhanced response to a stimulus which is normally painful. Pain induced by normally non-painful stimuli is termed *allodynia*. However simple this appears at first sight, there is no common opinion in the literature, when to use which term. Treede and co-authors [26] provide an insight into the history of these terms explaining this dilemma. Here we summarize it briefly, to explain possible discrepancies between our literature analysis and the original reports.

## *Hyperalgesia*

"Shortly after the detection of primary nociceptive afferents, studies noted that these increase their sensitivity upon repeated application of noxious stimuli. This phenomenon was called sensitization"(see [27]). *Peripheral sensitization*, or sensitization of primary afferents, fully accounts for thermal hyperalgesia that develops within the injured or inflamed skin area i.e. primary zone (*primary hyperalgesia*) and is mediated by A-fiber and C-fiber polymodal nociceptors. For some time, "hyperalgesia was seen as the perceptual counterpart of peripheral sensitization." Soon it became clear, that this concept "can not be generalized to other types of hyperalgesia... Enhanced responsiveness of nociceptive neurons in the spinal cord... (*central sensitization*) significantly contributes to several types of hyperalgesia", especially to those which are not restricted to the primary, but also spread onto a surrounding i.e. secondary zone (*secondary hyperalgesia*). Therefore, the authors propose to "define hyperalgesia as the perceptual counterpart of any sensitization, either in the peripheral or central nervous system".

## *Allodynia*

"While neurobiologists were working on the neural basis of primary hyperalgesia to heat, clinicians made observations on a puzzling phenomenon. Some patients, particularly after peripheral nerve lesions, experienced pain from gentle touch to their skin."

Soon it was established, that "this strange pain sensation is mediated by A-beta low-threshold mechanoreceptors (touch receptors) [28], [29]. A new term was introduced: *allodynia*. "Semantically, this term implies a neurophysiological mechanism, because it means pain elicited by a stimulus that is alien to the nociceptive system" (*allos* means *alien* and *odyne* means *pain* in Greek). Later, the term *allodynia* gradually moved away from its original clinical meaning. Now it is often applied to hypersensitivity to normally non-painful stimuli of any modality, whereas *hyperalgesia* is reserved for hypersensitivity to suprathreshold stimuli.

Treede and co-authors object to the physiological justification of the latter subdivision. The decisive question is: "Are threshold and suprathreshold changes two independent phenomena or are they linked together?" Experiments [30] have shown, that hyperalgesia includes both an increase in pain to suprathreshold stimuli and a decrease in pain threshold. This observation indicates that hyperalgesia and allodynia are just two aspects of the same phenomenon. Therefore, the authors [26] advocate for "the traditional use of *hyperalgesia* as an umbrella term for all phenomena of increased pain sensitivity" which does not imply any mechanisms. *Allodynia* may be used in cases, when it is clear, that sensitization happened to non-nociceptive afferents (A-beta-fibers). Many other researchers share this opinion, and so do we. From now on we will use *hyperalgesia* as the "umbrella term".

# Table 3A. Changes of activation (↑ - increases, ↓ - decreases) during hyperalgesia in the pain relevant structures of: A – human and B – rat brain

*Abbreviations of the brain structures* like in Figure 2. *Additional abbreviations*: e – experimental, c – clinical study, SMP - sympathetically mediated pain, IBS - irritable bowle syndrome, CRPS - complex regional pain syndrome, FM – fibromyalgia, CLBP - chronic low back pain, "S equ HC pain intensity" – stimulation (S) of the patients resulting in the same subjective pain intensity as in healthy controls (HC), "S equ HC pressure" – stimulation of the patients with the same pressure as the healthy controls, CCI - chronic constrictive injury. Note that at the place of *PremotC* at A there is *Z incer* - zona incerta - at B. Also *AIns* and *M-PIns* at A are replaced by *Amy* and *HTh* (see the text for further explanations).

## Table 3B

| Study | Hyperalgesia trigger | Imaging modality | BSt | Th | MTh | LTh | S1 | S2 | Pa | ACC PCC, PFC, RSC FC | Ins | HipF | Amy | HTh | Z incer | MotC | Cb |
|---|---|---|---|---|---|---|---|---|---|---|---|---|---|---|---|---|---|
| **Heat** | | | | | | | | | | | | | | | | | |
| Hess et al., 2007 | zymosan: S injected paw | fMRI | | | | | | | | | | | | | | | |
| Hess et al., 2007 | zymosan: S non-injected paw | fMRI | | | | | | | | | | | | | | | |
| Paulson et al., 2002 | CCI of sciatic nerve | rCBF ARG | | | | | | | | | | | | | | | |
| **Light brush** | | | | | | | | | | | | | | | | | |
| Malisza et al., 2003 | capsaicin i.i.: S at the paw | fMRI | | | | | | | | | | | | | | | |
| **Pin-prick** | | | | | | | | | | | | | | | | | |
| Moylan Governo et al., 2006 | capsaicin fM | RI | | | | | | | | | | | | | | | |
| Paulson et al., 2002 | CCI of sciatic nerve | rCBF ARG | | | | | | | | | | | | | | | |
| **Chemical** | | | | | | | | | | | | | | | | | |
| Porro et al., 1999 | -endorphin bklockade: S with formalin 2DG | ARG | | | | | | | | | | | | | | | |

## Ways of Sensitization

Sensitization, as the initial event of hyperalgesia, develops during pathophysiological states like inflammation or nerve injury. For investigations of sensitization mechanisms experimental models of inflammatory and of chronic neuropathic pain have been established. In the former ones, different irritants (e.g. capsaicin) are either subcutaneously injected or topically applied to the skin at peripheral body sites (e.g. limbs). These models are exploited in animal and in human studies. The latter models involve more severe damage of neural tissue (e.g. chronic constriction injury (CCI) of the sciatic nerve). They are only applicable in animal studies. Additional human-specific kind of sensitization can be provoked by psychological manipulations.

## Biases of Imaging Methods

For abbreviations of the brain structures in the text below please see the legend to Figure 2.

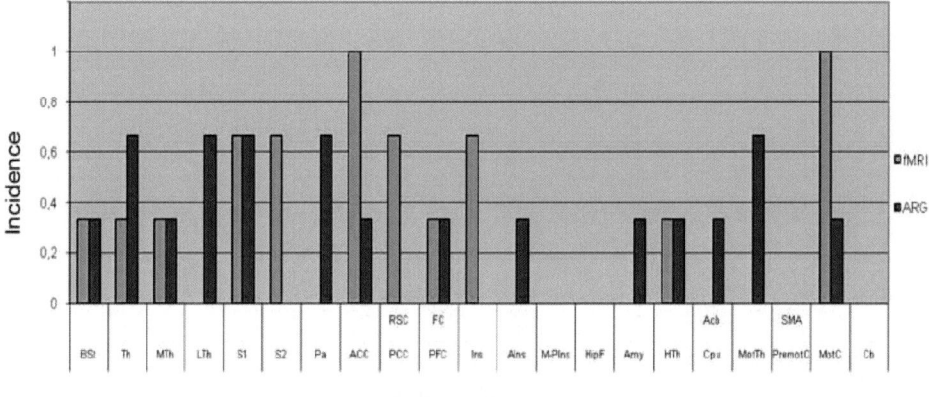

Explanation. BSt – brainstem. Some human studies defined the activated structures of the brainstem more precisely; the variants were: midbrain, trigeminal nucleus, PAG and superior colliculus. Rat autoradiography studies were capable of delineating single small nuclei of the brain stem reticular formation, different parts of PAG and pretectal area. These all are included here in BSt. MTh – medial thalamus, LTh – lateral thalamus. The cases where the thalamic activation was not determined by the authors as belonging to either medial or lateral thalamus are referred here as Th – thalamus. S1 – primary somatosensory cortex, S2 – secondary somatosensoy cortex, Pa – parietal cortices, including human parietal association cortex (Brodmann Areas 5/7) and also areas 40/39. ACC – anterior cingulate, PCC - posterior cingulate, RSC - retrosplenial, FC – frontal, PFC – prefrontal cortex. Many human studies delineated activations in different PFC zones; the variants were: dorsolateral prefrontal, orbitofrontal, superior, middle and inferior frontal cortex. These all are included here in PFC. AIns – anterior insula, M-PIns – middle and/or posterior insula; the cases in which the location was not specified are refered here as Ins – insula. HipF – hippocampal formation including entorhinal cortex, Amy – amygdala, HTh – hypothalamus, Cpu – caudate putamen, Acb - nucleus accumbens, MotTh – motor nuclei of the thalamus, PremotC – premotor cortex, SMA - supplementary motor area, MotC – (primary) motor cortex, Cb – cerebellum.

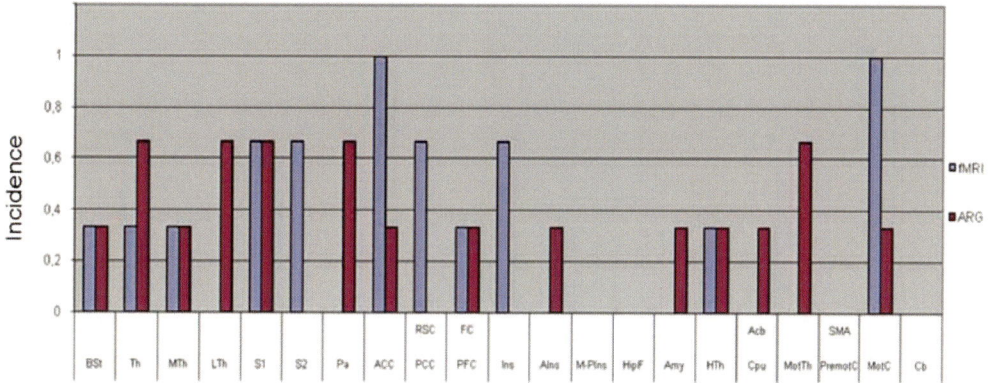

Figure 2. Incidence of the pain related brain structures showing increased activation under hyperalgesia (frequency/N° of studies) in: A - 23 fMRI and 10 PET human studies and B - 3 fMRI and 3 autoradiography (ARG) rat studies.

Due to their methodological peculiarities, different imaging modalities have certain biases in visualisation of different brain areas. For instance, MEG is more sensitive to the sources in *S1* and *S2* that are oriented tangentially to the scalp. EEG can better detect radially oriented current flows, so it shows more frequently the activity in *ACC*. At the same time, the both methods, although still unsurpassed in terms of temporal resolution, are hardly suitable for demonstrating activity in the deep brain structures like *Ins* or *Th*, and also in *PFC*. PET and fMRI do not have these drawbacks, still there is some evidence that PET can detect deeper structures better than fMRI. For example, Peyron and colleagues [31] in their combined PET/fMRI study of the same patient, were able to show activation in thalamus with PET, but not with fMRI. Autoradiographic imaging is still the best with respect to spatial resolution throughout the whole animal brain, but its temporal resolution is poor. Activations in response to short test stimuli cannot, strictly speaking, be detected with this method; they are always confounded by the ongoing activity (in our case pain). (This is the reason why we do not consider here autoradiography studies demonstrating rat brain activation in the chronic pain and hyperalgesia models, where no additional correlation analysis was performed between metabolic changes and behavioural hyperalgesia signs).

To estimate roughly the biases of imaging modalities in the studies under review, we calculated the incidence of the pain relevant brain structures, which showed increased activation during hyperalgesia, for different imaging modalities. 2 MSI works reported activation only in *S1*, 1 MEG – in *S1* and *ACC*, and 1 LEPs study – only in *ACC*; no other structures were mentioned. Because of their comparatively small number, we did not include these studies in Figure 2A, representing relative occurrence of brain structures in 23 fMRI and 10 PET human experiments. This figure shows that a relatively higher number of the PET studies detected activation in the deep structures, than did fMRI. On the contrary, cortical structures were better represented in the fMRI studies. Otherwise PET and fMRI studies were in agreement with each other. Figure 2B summarizes the data from 3 fMRI and 3 autoradiography rat studies under review. The number of these studies is insufficient to make final conclusions, but it is known, that animal fMRI was till recently more efficient in detecting activity in cortical areas, than in the deeper brain structures, and this trend can be recognized on Figure 2B. Besides, fMRI revealed an enhanced activation in the same structures, as autoradiography.

Altogether Figure 2 shows that almost all structures of the human or animal brain involved in pain processing show enhanced activity due to hyperalgesia, however to different degrees. This and the roles of structures relevant for hyperalgesia will be discussed 2 sections below.

## BRAIN STRUCTURES: HUMANS VERSUS RATS

Of course, functional parallels between certain human and rat brain structures cannot be set unambiguously. Especially this regards higher associative forebrain areas like *PFC*, which are, needless to say, much more developed in humans, than in rats. Many of the human studies under review delineated activations in different zones of *PFC* known to have different functions. Such delineation could not be done for rats. Thus, for simplicity of our comparative analysis, we had to consider all these zones under the general commonly used term *PFC* (Figure 2, Table 3). In the literature one can find "controversial views on the existence of *PFC* in non-primate mammals, in particular in rats" [32]. For our hyperalgesia work [22] we used the 4-th edition of Paxinos and Watson's Rat Brain Atlas [33]. According to it, we defined the anterior cortical area lying next to cingulate cortex as motor cortex. Uylings and colleagues in their study advancing the view that rats do possess *PFC* [32], provide evidence that this area functionally belongs to *PFC*. Malisza and co-authors [20] defined this area as frontal cortex. This heterogeneity in the rat frontal structures nomenclature may have led to an overestimation in rats of the motor cortex and underestimation of the *PFC* activity (Figure 2B). On the contrary, limbic *Amy* and *HTh* are much better detected in rats, possibly, due to anatomic specifics of the rat versus human brain.

### Deactivations

The nature of signal decreases under painful or hyperalgesic conditions in functional imaging is a matter of debate. However, these decreases may reflect actual reduction of neuronal and/or metabolic activity. The leading hypothesis regarding stimulus-evoked deactivations suggests that brain structures are at rest functionally interconnected in a "default mode" resulting in a structure-specific baseline activity. During a novel task, processing resources might be moved from some structures to others relevant for this task [34], [35]. This hypothesis explains deactivations observed in several hyperalgesia studies outside the pain matrix [36, 37, 38, 39]. However, some studies reported stimulus-evoked deactivations or lesser activations in patients than in healthy controls also within parts of the pain matrix: in *Th* [36, 40, 41], *PFC* [37, 39, 41, 42, 43], *ACC* [37, 42], *S2* and *Ins* [44] and *Cb* [36]. These studies, except [39], have been done on patients with different chronic disorders which have lasted for years. It is plausible, that during this time plastic (compensatory) changes have happened in pain processing circuits, which also influenced stimulus-evoked responses. For instance, unpleasantness, the affective component of pain (reflected in *ACC*) can be reduced by repeated stimulation, subject training, and habituation to the experimental paradigm. [3,

45, 46]. The same probably regards chronic pain states [31, 37]. At least, decreases in the thalamic and cortical areas by chronic pain [2, 8, 13, 40, 47, 48, 49] are well known.

The list of putative reasons of deactivations is very broad (cf. [50]) and could become a topic of a special review. At this point we would like to refer the reader to Table 3 representing all the changes, increases and decreases, observed by hyperalgesia, and concentrate on the brain structures showing increased activation.

## BRAIN STRUCTURES ACTIVATED DUE TO HYPERALGESIA

### Thermal Hyperalgesia

It is principally difficult to distinguish the network of structures activated solely by hyperalgesia from that activated by pain. The most elegant attempt has been undertaken by Lorenz and colleagues [51] who compared capsaicin-induced hyperalgesia with equally intense pain in the same subjects. The choice of thermal stimulation was also advantageous: Thermal hyperalgesia is known to be restricted to the primary zone (Tab. 2), and thus, one could expect it to be mediated by sensitized peripheral heat nociceptors activating the same central pathways as normal heat pain. The involvement of additional confounding factors (mechanical sensitization etc.) was this way excluded or minimizedl. If this so to speak "pure" hyperalgesia was mediated by the same network, as the equally intense pain, the brain activity during these two states would not have significantly differed. But this was not the case. The investigators found bilaterally increased activity by hyperalgesia in the following structures: *MTh*, *AIns*, ventral *Cpu*, perigenual (Brodmann Area (BA) 32) *ACC*, orbitofrontal (OFC; BA 10/11) and dorsolateral prefrontal (DLPFC; BA 9/8) cortex, dorsomedian midbrain (Table 3A). Primary sensory-motor and premotor cortices, and bilateral *Cb* responded during heat pain (see also [52]), but not during hyperalgesia.

In other words, this comparison revealed specific activation by heat hyperalgesia of the medial thalamic pathway to the frontal lobe and thus the medial pain system (Figure 1). Ventral *Cpu* belongs to ventral associative-limbic loop, that includes *MTh*, OFC, perigenual *ACC* and *AIns* [53], forming a link to the motor system influenced by motivational context. Ventral striatum and OFC play a salient role in pain/aversion processing or anticipation [54, 55]. *AIns* and the other limbic structures have been shown to get activated by distressing emotions, and medial *PFC* and *Th* - by any emotions [56]. Ploghaus and co-workers [57] have shown, that thermal hyperalgesia induced only by anxiety is associated with activity increases in the entorhinal cortex of *HipF* and affective perigenual cingulate (in this case BAs 24/32) cortex. No changes in *Th* or *S1* have been reported. Participation of medial *PFC* (BA 11) in affect processing is endorsed by the findings of Wiech and colleagues [58]. Taken together, the medial pain system and limbic structures are associated with emotional, motivational and cognitive aspects of pain [59] (also Figure1). Hence, the simplest functional explanation for the peculiar pattern of brain activity by heat hyperalgesia would be a greater unpleasantness caused by hyperalgesic as compared to normal pain.

However, Lorenz and colleagues [51] (see also [60]) emphasize "factors other than unpleasantness... *PFC* is activated consistently when painful experiences occur in the context of tissue alterations resulting from capsaicin treatment (this study, [46, 61]), in response to trauma [62]... or neuropathic pain [49]... The prefrontal and orbitofrontal cortical activity during this conditions probably reflects cognitive and emotional responses to perceived tissue pathology." This way "the brain recognizes the unique physiological features of different painful conditions, thus permitting adaptive responses" to them.

Maihoefner and Handwerker [63] obtained fMRI brain activation patterns during thermal and during pin-prick (see *Pin-prick*) hyperalgesia adjusted to result in equal pain intensity ratings after capsaicin application. Moreover, subjects had to rate the pain unpleasantness. Brain areas activated during thermal hyperalgesia (i.e. subtraction of brain activations during thermal stimulation before and after capsaicin exposure) were *S1* and *S2*, *Ins*, associative somatosensory cortices (BA 5/7, 40), gyrus cingulum (GC; BAs 24' and 32') and also superior (SFC; BA 6/8), middle (MFC; BA 9/10) and inferior frontal cortex (IFC; BAs 44/45). Similarly to [51], this study demonstrated increased activity in insular, cingulate and expanded *PFC* areas. Contrary to [51], it also demonstrated considerable increases in somatosensory areas. Here, as distinct from [51], the stimulation temperature before and after capsaicin exposure was the same. Whereas Lorenz and co-authors compared hyperalgesia with equally intense pain, Maihoefner and Handwerker compared hyperalgesia with the feeling of warmth which implicates involvement of additional peripheral nociceptors during hyperalgesia and stronger recruitment of those already activated by warmth. This could explain the additional activation in *S1*, *S2* and consequently in the associative somatosensory cortices. The same concerns our fMRI study of zymosan-induced hyperalgesia in rats [22]. In this study we also compared normal versus hyperalgesic pain at the same stimulus intensity thereby confirming an enhanced activation of somatosensory cortex. Nevertheless we also found prominent increases within the medial pain system (*MTh*, cingulo-frontal and insular areas), in motor (*MotC*) and limbic (*HTh*) structures (Table 3B).

A method of [99m]Tc-exametazime autoradiography was developed in the laboratory of K. Casey [64], [1] permitting analysis of the regional cerebral blood flows (rCBFs) much sooner after stimulation (so to speak, having a higher temporal resolution than the 2DG technique),. Using this method the researchers investigated forebrain activity in rats with sciatic nerve CCI [65, 66]. In [66] they correlated different behavioural changes with the forebrain activity. High correlation with heat hyperalgesia was found in somatosensory cortex, specifically in HL and parietal areas, and in *HTh* and *Amy*. Again, the enhanced activation in *S1* could have originated from the enhanced recruitment at the periphery. Furthermore, one should not forget that autoradiographic findings always comprise some baseline brain activity (see *Biases*), which in this case reflected the ongoing pain, like in the other autoradiography studies reporting hyperalgesia [65, 67, 68].

In the second part of the study by Maihoefner and Handwerker [63], a comparison of brain activation patterns during thermal and pin-prick hyperalgesia has been made. It revealed a significantly greater activation during thermal hyperalgesia of contralateral SFC, IFC and of bilateral GC, MFC and *AIns*. The absence of differences in the regions involved in pain intensity coding, *S1* or *S2*, reflected the similar pain intensity ratings for both hyperalgesia types. Stronger activations of GC, MFC and *AIns* significantly correlated with higher ratings of the heat pain-related unpleasantness. These findings, a shift towards the medial pain system and, correspondently, an increased pain unpleasantness by thermal hyperalgesia are, again, in agreement with the results of Lorenz and colleagues [51, 69]. The authors [63] also emphasize the role of the frontal lobes in cognitive, attentional and motor-related processing and consider these functions especially essential during hyperalgesia.

6 other studies under review investigating thermal hyperalgesia or abnormal processing of thermal pain at somatic sites (hand or face) were performed on patients with chronic disorders: atypical facial pain [43], rheumatoid arthritis [42], sympathetically mediated pain [40], irritable bowle syndrome [70], migraine [71] and burning mouth disorder [36]. In spite of extremely different origin and mechanisms underlying these disorders, one still can find certain common features in these works, and in studies on experimental hyperalgesia. Apart from deactivations or activations lesser than in healthy controls, only one work reported increased activity during hyperalgesia in *MTh*, *S1*, *S2* and *Ins* [70], one work – in parietal association cortex [36], two studies – in *PFC* and 5 – in *ACC* (BAs 24/32). Apparently, the stimulated sites in these studies were not necessarily or not at all affected by the disease. This can probably explain the lack of general increases in thalamus and consequently in somatosensory cortex: there was not much or no additional input from the periphery, as compared to stimulation of healthy controls / in chronic pain alleviated states. The processes of sensitization have to occur probably at higher processing levels, so that hyperalgesia in these cases showed up only in cingulo-frontal networks. This resembles the situation observed in our rat zymosan model [22] during stimulation of the non-injected paw: under hyperalgesia conditions significant blood oxygenation level dependent (BOLD) signal increases have been detected only in higher associative (cingulo-frontal, insular) and output-related (MotC) brain structures; we named this phenomenon "cerebral sensitization".

## Mechanical Hyperalgesia

### *Light Touch*

Hyperalgesia to light stroking or brushing (dynamic hyperalgesia) is obviously the most interesting for the clinics. It implies activation of the other receptors, originally pain insensitive A-beta-mechanoreceptors (Tab. 2) i.e. allodynia. One of the first investigations of capsaicin-induced brush-evoked allodynia has been performed by Iadarola and colleagues [61] with PET. Subtraction of the scans acquired during light brushing from those during allodynia revealed residual activations in: the midline of ventral midbrain; lateral *Cb*, ipsilaterally; superior frontal gyrus (BA 10) more contra- but also ipsilaterally; parahippocampal gyrus contralaterally; *Ins* bilaterally; cingulate gyrus, inferior parietal lobule (IPL)/*S2* bilaterally; *S1* contralaterally. Two regional activations were particularly prominent by hyperalgesia firstly in *PFC*, with the focus within BA 10, but extending also to BAs 9 and 11, secondly in parahippocampal gyrus. The authors conclude that this activation "contains a cognitive and limbic component, either not present or inconsistently present, during acute pain or light tactile stimulation." Maihoefner and colleagues [72] used a similar experimental model and fMRI. They found out, that brushing the allodynic skin produced significant BOLD signal increases, as compared to non-painful control stimulation of the other arm, in the contralateral *PFC* (BA 9), *Pa* and bilaterally in *Ins*, and also in the contralateral *S1* and bilateral *S2*.

The other imaging studies of dynamic hyperalgesia confirmed the importance of higher associative structures: the middle (BAs 6, 8, and 9) and inferior frontal gyrus (BAs 44 and 45) [46] or sensory association cortex (BA 5/7), *PFC* (BA 9/10/47) and *Ins* [73], but reported no increases of activity in *S1*. The absence of increases in *S1* during hyperalgesia in [46] could probably be explained by the lower pain intensity rating than e.g. in [61]. A later study from this group [74] was performed with MSI and aimed at measuring activation in *SI* during brush-evoked hyperalgesia. Somatosensory evoked magnetic fields (SEFs) were induced by non-painful electrical stimulation of A-beta-afferents first in the normal skin and then after capsaicin injection. During A-beta-hyperalgesia the SEF amplitudes were significantly higher. The discrepancy with the previous study [46] the authors ascribe to a higher sensitivity of MSI, particularly in *SI* (see *Biases*).

The hyperalgesia investigations by the group of D. Borsook deserve special attention. They researchers succeeded in imaging the whole trigeminal sensory system, from trigeminal ganglion through the trigeminal nucleus and thalamus to *SI*, with fMRI [75, 76]. Further, they investigated heat and brush-evoked allodynia following heat/capsaicin sensitization of the face skin [77]. Increased fMRI signals were observed in both cases in the ipsilateral trigeminal nucleus, and by brush-induced allodynia also in the contralateral *LTh* (VPM). Obviously the fact, that the face has a much larger somatotopic representation in the sensory thalamus than forearms or lower legs - the usual stimulation sites - permitted excellent measurement also in thalamus. Only 1 of the 10 other studies of light touch hyperalgesia under review reported increases of thalamic activity [78], at the same time 7 of them detected increases in somatosensory cortex (Tab. 3A). It seems unlikely, that *SI* in these studies has been activated only via indirect influences, without involvement of thalamic structures. Rather, the thalamic activity has been underestimated.

Malisza and colleagues [20] have introduced the capsaicin model into the rat fMRI. Light touch allodynia at the paw following capsaicin injection into the ankle joint manifested itself in *ACC* bilaterally; frontal bilaterally; and sensory-motor cortex contralaterally.

The 6 studies on chronic pain patients endorsed the importance of cingulate (all, including [79]), prefrontal [38, 80, 81], parietal association [38, 44, 78], insular [38, 44, 80] and higher motor and motor [38, 44] cortices for brush-evoked hyperalgesia processing. Most of them also reported activity increases within *SI*. However, Witting and colleagues [80] detected rCBF increases in *SI* only during brushing of unaffected, but not of allodynic skin. The authors suppose that "the absence of *SI* activity is pathophysiologically linked to allodynia" and conclude: "…activity in the sensory-discriminative network is downregulated in neuropathic pain. Instead, there is an upregulation of activity in OFC and insula probably due to a stronger emotional load of neuropathic pain and higher computational demands of processing a mixed sensation of brush and pain."

In summary, as distinct from the cases of "pure" thermal hyperalgesia, dynamic mechanical hyperalgesia causes additional increases of activation within the sensory-discriminative lateral pain system (VPM, *S1*, *S2*). But since we know, that allodynia implies activation of additional receptors, A-beta-mechanoreceptors, this difference is not surprising. Nevertheless, similarly to thermal hyperalgesia, one can see here a shift of activation emphasis towards the structures responsible for cognitive and emotional evaluation of pain.

### *Cold Rubbing*

The mechanism of hyperalgesia evoked by cold stimuli is less understood (cf. Table 2). It has been noticed however, that light touch allodynia can be enhanced by cold. In 3 studies under review by the research group of R. Peyron [31, 37, 44], innoxious movement of a cold subject on the skin has been used as the test stimulus. These studies have been performed on allodynic pain patients with lesions in different parts of the nervous system. Over-activity of the lateral thalamus and, consequently, increases in *S1*, *S2*/IPL and *AIns* have been admitted as remarkable features of this allodynic pain processing. The authors emphasize responses in the ipsilateral *S1*, *S2* and *Ins* as the most salient effect of allodynia [44], (see also [8]). Laterality of responses in the human brain is a special complicated issue on which we do not focus in the present review. Nevertheless, we should admit, that spread of activation onto the ipsilateral structures under hyperalgesia conditions has been demonstrated by other human fMRI, e.g. [72, 82], and rat autoradiography studies. In our rat fMRI study [22] we also observed signal increases in the ipsilateral *S1*, *S2* and *Ins* during hyperalgesia.

*Pin-Prick*

Pin-prick (punctate) hyperalgesia is mediated predominantly by A-delta-nociceptors and does not implicate additional afferents. However, they have an extended and dens terminal arborization and high discharge rates [83] probably resulting in a spread of activity i.e. the secondary zone. A brief view at Tab. 3A convinces: pin-prick hyperalgesia activates the lateral pain system too. From the 1 PET and 4 fMRI studies using experimental models and 1 fMRI study on CRPS patients (altogether 5), 3 studies reported enhanced activity in *Th* [39, 45, 80, 84]. In particular, Kupers and co-authors [80, 84] investigated hyperalgesia at the face skin and managed to delimit activity increases in *LTh* (VPM), as well as in *MTh*. 3 studies detected increases in *S1*, *S2* and somatosensory-association cortices [45, 63, 85, 86]. One more study on CRPS patients was done with MSI [86]. It also demonstrated increases in *S1*.

However, Table 3A also demonstrates amplified recruitment during pin-prick hyperalgesia of the cingulate [39, 45, 80, 84, 85], prefrontal and premotor [45, 63, 85] cortices, thus confirming affective and cognitive accentuation also for this hyperalgesia form.

The study by Zambreanu and colleagues [45] additionally concentrated on brainstem activation by hyperalgesia. It was localised to two distinct midbrain areas – nucleus cuneiformis (NCF) and superior colliculi/periaqueductal gray (SC/PAG). PAG and NCF are the major sources of input to the rostral ventromedial medulla, which is presently considered as the final common output for descending influences (for review see [87]). The authors [45] suggest that NCF and PAG are indeed involved in central sensitisation in humans. They suppose that such phylogenetically ancient structure as the brainstem reticular formation plays an important role in mediating hyperalgesia. A bit contradictory to these findings are the results of Moylan Governo and co-workers [21] who detected negative BOLD signal in PAG during pin-prick stimulation of capsaicin-treated rats. The physiological meaning of negative BOLD responses is unclear. For these two studies it is unclear, whether the activated mesencephalic neurons were excitatory on-cells or inhibitory off-cells driving protective descending modulation (see [88]). Anyway, both the human and the rat study endorsed important role of PAG and other *BSt* areas in processing of hyperalgesia.

*Pressure*

In contrast to pin-prick, hyperalgesia to blunt pressure is mediated by mechanically-insensitive C nociceptors [89] and restricted to primary zone (Table 2). So, like in the case of thermal hyperalgesia, one could theoretically expect pressure hyperalgesia to activate the same central networks as normal pressure pain.

Gracely and colleagues [41] and later Giesecke and colleagues [82] addressed a question, whether patients with idiopathic chronic disorders like fibromyalgia or chronic low back pain (CLBP) exhibit generalized increased pain sensitivity, hyperalgesia. They applied noxious pressure at a neutral site (thumbnail) and firstly determined the individual pain sensitivities.

Then, both performed two fMRI measurements: one with equal pressure, and the other with equal individual pain rating for healthy controls and patients. The both works revealed hyperalgesia in CLBP [82] and in fibromyalgia [41], [82] groups. In [82] stimulation evoking equal pain, resulted in common areas of activations in all 3 groups (CLBP, fibromyalgia and healthy controls): in the contralateral *S1*, *S2*, IPL, *Ins* and *ACC*, and ipsilateral *S2* and *Cb*. Stimulation with equal pressure resulted in moderate pain in the patients and just faint pain in the controls. fMRI detected only a single activation in controls - the contralateral *S2*. 5 common regions of neuronal activation were found in the CLBP and fibromyalgia groups: the contralateral *S1*, S2 and IPL, *Cb*, and ipsilateral *S2*. Additionally, Gracely and colleagues [41] using equal pressure stimuli reported stronger activation in *ACC*, *PCC* and *Ins* in fibromyalgia patients. These findings corroborate their hypothesis of central pain augmentation in chronic pain patients. In the case of pressure hyperalgesia, it manifested itself in the cortical networks mediating normal pressure pain.

*Chemical Stimulation*

Besides capsaicin and zymosan, mainly applied to evoke sensitisation as discussed above, formalin is also a widely used irritant. Although, the classical formalin test is appropriate for estimation of acute tonic pain, many researchers consider it less appropriate for studying hyperalgesia, therefore here we do not consider a few imaging studies which "visualized" formalin-induced pain only. Porro and colleagues [90] investigated with 2DG autoradiography formalin pain in rats on the background of partial β-endorphin system blockade. The rats exhibited hyperalgesia – increased licking response. Correlation between the licking response duration and metabolic activity was calculated. The structures with the high correlation coefficient (>0.65) were: anterior pretectal nucleus, zona incerta (involved in sensory-motor integration and antinociception), parafascicular, centrolateral, posteromedial, ventroposterolateral, reticular and ventrolateral thalamic nuclei, posterior parietal cortex, parietal area 1 and frontal areas 1, 2, 3. The authors drew a conclusion: "Frontolimbic networks may be critically involved in shaping pain-related behaviour" during hyperalgesia.

## COMMENTS AND CONCLUSION

As we have seen, state-of-the-art imaging studies addressing hyperalgesia, as a distinct from normal pain phenomenon, are not numerous, but very heterogeneous. Since the main 3d imaging methods (PET and fMRI) exhibit just quantitative, but not qualitative differences in their capability to detect signals from the different brain areas, the differences between the findings of the studies under review cannot be ascribed to methodological biases. Rather, they are associated the different forms of hyperalgesia studied. Nevertheless, our general impression is that these different forms have something in common, namely: a "shift" of the pain processing emphasis towards the medial pain system and a prominent involvement of affective-motivational, cognitive and limbic circuits. This is especially evident in the case of thermal hyperalgesia devoid of any mechanical component. Remarkable in this respect are the findings of Ushida and co-authors [91], who evoked hyperalgesia in allodynia patients without any "physical" manipulation. They directly "stimulated their emotions" by a movie showing brushing of their allodynic hands, and detected activations in *ACC* (BA 24) and *PFC* (BA 10).

More importantly, we asked here the question, whether SAI can contribute to our understanding of this clinical syndrome. Now we answer: yes. The rat hyperalgesia studies generally supported the findings of the human studies. Even though only few *in-vivo* functional imaging studies of hyperalgesia in rats have been performed till now, we can say, that they depict stimulus-induced hyperalgesia more appropriately, then autoradiography, due to the higher temporal resolution. fMRI, mainly due to its high spatial resolution and complete non-invasiveness, proves to be currently the best imaging technique, because the very same method used in preclinical research on animals can be transferred to the clinics. The latter and the high quality of the forthcoming small animal fMRI should also permit to reduce the amount of experimental animals and therefore contribute to animal welfare.

## ACKNOWLEDGMENTS

For financial support of our research we would like to thank Doerenkamp Professorship in Innovations in Animal and Consumer Protection, the SET-Stiftung, the DFG FG-661 TP4 and the BMBF Migraine research group 01 EM 0514.

## REFERENCES

[1] Casey KL. Forebrain mechanisms of nociception and pain: analysis through imaging. *Proc. Natl. Acad. Sci. USA* 1999; 96: 7668-7674.

[2] Derbyshire SW. Meta-Analysis of Thirty-Four Independent Samples Studied Using PET Reveals a Significantly Attenuated Central Response to Noxious Stimulation in Clinical Pain Patients. *Curr. Rev. Pain* 1999; 3: 265-280.

[3] Disbrow E, Baron R ,Baron Y. Examination of the Role of the Cerebral Cortex in the Perception of Pain Using Functional Magnetic Resonance Imaging. *Curr. Rev. Pain* 1999; 3: 281-290.

[4] Pawl RP. A Review of Functional Imaging of the Brain and Pain. *Curr. Rev. Pain* 1999; 3: 249-255.

[5] Treede RD, Kenshalo DR, Gracely RH, et al. The cortical representation of pain. *Pain* 1999; 79: 105-111.

[6] Casey KL. Concepts of pain mechanisms: the contribution of functional imaging of the human brain. *Prog. Brain Res.* 2000; 129: 277-287.

[7] Hudson AJ. Pain perception and response: central nervous system mechanisms. *Can. J. Neurol. Sci.* 2000; 27: 2-16.

[8] Peyron R, Laurent B ,Garcia-Larrea L. Functional imaging of brain responses to pain. A review and meta-analysis (2000). *Neurophysiol Clin.* 2000a; 30: 263-288.

[9] Rainville P, Bushnell MC ,Duncan GH. Representation of acute and persistent pain in the human CNS: potential implications for chemical intolerance. *Ann. NY Acad. Sci.* 2001; 933: 130-141.

[10] Borsook D ,Becerra L. Pain imaging: future applications to integrative clinical and

basic neurobiology. *Adv. Drug Deliv. Rev.* 2003; 55: 967-986.

[11] Porro CA. Functional imaging and pain: behavior, perception, and modulation. *Neuroscientist* 2003; 9: 354-369.

[12] Verne GN, Robinson ME, Price DD. Representations of pain in the brain. *Curr. Rheumatol. Rep.* 2004; 6: 261-265.

[13] Apkarian AV, Bushnell MC, Treede RD, et al. Human brain mechanisms of pain perception and regulation in health and disease. *Eur. J. Pain* 2005; 9: 463-484.

[14] Borsook D, Ploghaus A, Becerra L. Utilizing brain imaging for analgesic drug development. *Curr. Opin. Investig. Drugs* 2002; 3: 1342-1347.

[15] Wise RG, Tracey I. The role of fMRI in drug discovery. *J. Magn. Reson. Imaging* 2006; 23: 862-876.

[16] Tuor UI, Malisza K, Foniok T, et al. Functional magnetic resonance imaging in rats subjected to intense electrical and noxious chemical stimulation of the forepaw. *Pain* 2000; 87: 315-324.

[17] Chang C, Shyu BC. A fMRI study of brain activations during non-noxious and noxious electrical stimulation of the sciatic nerve of rats. *Brain Res.* 2001; 897: 71-81.

[18] Malisza KL, Docherty JC. Capsaicin as a source for painful stimulation in functional MRI. *J. Magn. Reson. Imaging* 2001; 14: 341-347.

[19] Shah YB, Haynes L, Prior MJ, et al. Functional magnetic resonance imaging studies of opioid receptor-mediated modulation of noxious-evoked BOLD contrast in rats. *Psychopharmacology* (Berl) 2005; 180: 761-773.

[20] Malisza KL, Gregorash L, Turner A, et al. Functional MRI involving painful stimulation of the ankle and the effect of physiotherapy joint mobilization. *Magn. Reson Imaging* 2003a; 21: 489-496.

[21] Moylan Governo RJ, Morris PG, Prior MJ, et al. Capsaicin-evoked brain activation and central sensitization in anaesthetised rats: a functional magnetic resonance imaging study. *Pain* 2006; 126: 35-45.

[22] Hess A, Sergejeva M, Budinsky L, et al. Imaging of hyperalgesia in rats by functional MRI. *Eur. J. Pain* 2007; 11: 109-119.

[23] Meller ST, Gebhart GF. Intraplantar zymosan as a reliable, quantifiable model of thermal and mechanical hyperalgesia in the rat. *Eur. J. Pain* 1997; 1: 43-52.

[24] Negus SS, Vanderah TW, Brandt MR, et al. Preclinical assessment of candidate analgesic drugs: recent advances and future challenges. *J. Pharmacol. Exp. Ther.* 2006; 319: 507-514.

[25] Merskey H, and Bogduk, N. Classification of Chronic Pain: Descriptions of Chronic Pain Syndromes and Definitions of Pain Terms *IASP Press*, Seattle, 1994.

[26] Treede RD, Handwerker, H.O., Baumgärtner, U., Meyer, R.A. and Magerl, W. Hyperalgesia and allodynia: Taxonomy, Assessment, and Mechanisms. In: Brune, K, and Handwerker, H.O., Eds. Hyperalgesia: Molecular Mechanisms and Clinical Implications. 30. *IASP Press*, Seattle, 2004; 3-15.

[27] Woolf CJ, Salter MW. Neuronal plasticity: increasing the gain in pain. *Science* 2000; 288: 1765-1769.

[28] Gracely RH, Lynch SA, Bennett GJ. Painful neuropathy: altered central processing maintained dynamically by peripheral input. *Pain* 1992; 51: 175-194.

[29] Torebjork HE, Lundberg LE, LaMotte RH. Central changes in processing of mechanoreceptive input in capsaicin-induced secondary hyperalgesia in humans. *J.*

*Physiol.* 1992; 448: 765-780.

[30] Baumgartner U, Magerl W, Klein T, et al. Neurogenic hyperalgesia versus painful hypoalgesia: two distinct mechanisms of neuropathic pain. *Pain* 2002; 96: 141-151.

[31] Peyron R, Garcia-Larrea L, Gregoire MC, et al. Parietal and cingulate processes in central pain. A combined positron emission tomography (PET) and functional magnetic resonance imaging (fMRI) study of an unusual case. *Pain* 2000b; 84: 77-87.

[32] Uylings HB, Groenewegen HJ ,Kolb B. Do rats have a prefrontal cortex? *Behav. Brain Res.* 2003; 146: 3-17.

[33] Paxinos G, and Watson, C. *The rat brain in stereotaxic coordinates*. 4th ed. Academic Press, New York,1998.

[34] Gusnard DA, Raichle ME ,Raichle ME. Searching for a baseline: functional imaging and the resting human brain. *Nat. Rev. Neurosci.* 2001; 2: 685-694.

[35] Raichle ME, MacLeod AM, Snyder AZ, et al. A default mode of brain function. *Proc. Natl. Acad. Sci. USA* 2001; 98: 676-682.

[36] Albuquerque RJ, de Leeuw R, Carlson CR, et al. Cerebral activation during thermal stimulation of patients who have burning mouth disorder: an fMRI study. *Pain* 2006; 122: 223-234.

[37] Peyron R, Garcia-Larrea L, Gregoire MC, et al. Allodynia after lateral-medullary (Wallenberg) infarct. A PET study. *Brain* 1998; 121 ( Pt 2): 345-356.

[38] Maihofner C, Handwerker HO ,Birklein F. Functional imaging of allodynia in complex regional pain syndrome. *Neurology* 2006; 66: 711-717.

[39] Iannetti GD, Zambreanu L, Wise RG, et al. Pharmacological modulation of pain-related brain activity during normal and central sensitization states in humans. *Proc. Natl. Acad. Sci. USA* 2005; 102: 18195-18200.

[40] Apkarian AV, Thomas PS, Krauss BR, et al. Prefrontal cortical hyperactivity in patients with sympathetically mediated chronic pain. *Neurosci. Lett.* 2001; 311: 193-197.

[41] Gracely RH, Petzke, F., Wolf, J.M. and Clauw, D.J. Functional magnetic resonance imaging evidence of augmented pain pocessing in fibromyalgia. *Arthritis Rheum.* 2002; 46: 1333-1343.

[42] Jones AK ,Derbyshire SW. Reduced cortical responses to noxious heat in patients with rheumatoid arthritis. *Ann. Rheum. Dis.* 1997; 56: 601-607.

[43] Derbyshire SW, Jones AK, Devani P, et al. Cerebral responses to pain in patients with atypical facial pain measured by positron emission tomography. *J. Neurol. Neurosurg. Psychiatry* 1994; 57: 1166-1172.

[44] Peyron R, Schneider F, Faillenot I, et al. An fMRI study of cortical representation of mechanical allodynia in patients with neuropathic pain. *Neurology* 2004; 63: 1838-1846.

[45] Zambreanu L, Wise RG, Brooks JC, et al. A role for the brainstem in central sensitisation in humans. Evidence from functional magnetic resonance imaging. *Pain* 2005; 114: 397-407.

[46] Baron R, Baron Y, Disbrow E, et al. Brain processing of capsaicin-induced secondary hyperalgesia: a functional MRI study. *Neurology* 1999; 53: 548-557.

[47] Baron R. Mechanistic and clinical aspects of complex regional pain syndrome (CRPS). *Novartis Found Symp 2004*; 261: 220-233; discussion 233-228, 256-261.

[48] Iadarola MJ, Max MB, Berman KF, et al. Unilateral decrease in thalamic activity observed with positron emission tomography in patients with chronic neuropathic pain.

*Pain* 1995; 63: 55-64.

[49] Hsieh JC, Belfrage M, Stone-Elander S, et al. Central representation of chronic ongoing neuropathic pain studied by positron emission tomography. *Pain* 1995; 63: 225-236.

[50] Derbyshire SW. Burning questions about the brain in pain. *Pain* 2006; 122: 217-218.

[51] Lorenz J, Cross D, Minoshima S, et al. A unique representation of heat allodynia in the human brain. *Neuron. 2002*; 35: 383-393.

[52] Casey KL, Morrow TJ, Lorenz J, et al. Temporal and spatial dynamics of human forebrain activity during heat pain: analysis by positron emission tomography. *J. Neurophysiol.* 2001; 85: 951-959.

[53] Nakano K, Kayahara T, Tsutsumi T, et al. Neural circuits and functional organization of the striatum. *J. Neurol.* 2000; 247 Suppl 5: V1-15.

[54] Becerra L, Breiter HC, Wise R, et al. Reward circuitry activation by noxious thermal stimuli. *Neuron.* 2001; 32: 927-946.

[55] Jensen J, McIntosh AR, Crawley AP, et al. Direct activation of the ventral striatum in anticipation of aversive stimuli. *Neuron* 2003; 40: 1251-1257.

[56] Reiman EM, Lane RD, Ahern GL, et al. Neuroanatomical correlates of externally and internally generated human emotion. *Am. J. Psychiatry* 1997; 154: 918-925.

[57] Ploghaus A, Narain C, Beckmann CF, et al. Exacerbation of pain by anxiety is associated with activity in a hippocampal network. *J. Neurosci.* 2001; 21: 9896-9903.

[58] Wiech K, Seymour B, Kalisch R, et al. Modulation of pain processing in hyperalgesia by cognitive demand. *Neuroimage* 2005; 27: 59-69.

[59] Melzack R, and Casey, K.L. Sensory, motivational, and central control determinants of pain. In: Kenshalo, DR, Eds. *The Skin Senses*. Thomas, Springfield, IL, 1968; 423-443.

[60] Casey KL, Lorenz J ,Minoshima S. Insights into the pathophysiology of neuropathic pain through functional brain imaging. *Exp. Neurol.* 2003; 184 Suppl 1: S80-88.

[61] Iadarola MJ, Berman KF, Zeffiro TA, et al. Neural activation during acute capsaicin-evoked pain and allodynia assessed with PET. *Brain* 1998; 121 ( Pt 5): 931-947.

[62] Hsieh JC, Stahle-Backdahl M, Hagermark O, et al. Traumatic nociceptive pain activates the hypothalamus and the periaqueductal gray: a positron emission tomography study. *Pain* 1996; 64: 303-314.

[63] Maihofner C ,Handwerker HO. Differential coding of hyperalgesia in the human brain: a functional MRI study. *Neuroimage* 2005; 28: 996-1006.

[64] Morrow TJ, Paulson PE, Danneman PJ, et al. Regional changes in forebrain activation during the early and late phase of formalin nociception: analysis using cerebral blood flow in the rat. *Pain* 1998; 75: 355-365.

[65] Paulson PE, Morrow TJ ,Casey KL. Bilateral behavioral and regional cerebral blood flow changes during painful peripheral mononeuropathy in the rat. *Pain* 2000; 84: 233-245.

[66] Paulson PE, Casey KL ,Morrow TJ. Long-term changes in behavior and regional cerebral blood flow associated with painful peripheral mononeuropathy in the rat. *Pain* 2002; 95: 31-40.

[67] Mao J, Mayer DJ ,Price DD. Patterns of increased brain activity indicative of pain in a rat model of peripheral mononeuropathy. *J. Neurosci.* 1993; 13: 2689-2702.

[68] Neto FL, Schadrack J, Ableitner A, et al. Supraspinal metabolic activity changes in the rat during adjuvant monoarthritis. *Neuroscience* 1999; 94: 607-621.

[69] Lorenz J, Minoshima S ,Casey KL. Keeping pain out of mind: the role of the

dorsolateral prefrontal cortex in pain modulation. *Brain* 2003; 126: 1079-1091.

[70] Verne GN, Himes NC, Robinson ME, et al. Central representation of visceral and cutaneous hypersensitivity in the irritable bowel syndrome. *Pain* 2003; 103: 99-110.

[71] de Tommaso M, Guido, M., Libro, G., Losito, L., Difruscolo, O., Puca, F., Specchio, L.M. and Carella, A. Topographic and dipolar analysis of laser-evoked potentials during migraine attack. *Headache* 2004; 44: 947-960.

[72] Maihofner C, Schmelz M, Forster C, et al. Neural activation during experimental allodynia: a functional magnetic resonance imaging study. *Eur. J. Neurosci.* 2004; 19: 3211-3218.

[73] Witting N, Kupers RC, Svensson P, et al. Experimental brush-evoked allodynia activates posterior parietal cortex. *Neurology* 2001; 57: 1817-1824.

[74] Baron R, Baron Y, Disbrow E, et al. Activation of the somatosensory cortex during Abeta-fiber mediated hyperalgesia. A MSI study. *Brain Res.* 2000; 871: 75-82.

[75] DaSilva AF, Becerra L, Makris N, et al. Somatotopic activation in the human trigeminal pain pathway. *J. Neurosci.* 2002; 22: 8183-8192.

[76] Borsook D, DaSilva AF, Ploghaus A, et al. Specific and somatotopic functional magnetic resonance imaging activation in the trigeminal ganglion by brush and noxious heat. *J. Neurosci.* 2003; 23: 7897-7903.

[77] Borsook D, Burstein R ,Becerra L. Functional imaging of the human trigeminal system: opportunities for new insights into pain processing in health and disease. *J. Neurobiol.* 2004; 61: 107-125.

[78] Petrovic P, Ingvar M, Stone-Elander S, et al. A PET activation study of dynamic mechanical allodynia in patients with mononeuropathy. *Pain* 1999; 83: 459-470.

[79] Maihofner C, Neundorfer B, Stefan H, et al. Cortical processing of brush-evoked allodynia. *Neuroreport* 2003a; 14: 785-789.

[80] Witting N, Kupers RC, Svensson P, et al. A PET activation study of brush-evoked allodynia in patients with nerve injury pain. *Pain* 2006; 120: 145-154.

[81] Olausson H, Marchand S, Bittar RG, et al. Central pain in a hemispherectomized patient. *Eur. J. Pain* 2001; 5: 209-217.

[82] Giesecke T, Gracely RH, Grant MA, et al. Evidence of augmented central pain processing in idiopathic chronic low back pain. *Arthritis Rheum.* 2004; 50: 613-623.

[83] Ziegler EA, Magerl W, Meyer RA, et al. Secondary hyperalgesia to punctate mechanical stimuli. Central sensitization to A-fibre nociceptor input. *Brain* 1999; 122 ( Pt 12): 2245-2257.

[84] Kupers RC, Svensson P ,Jensen TS. Central representation of muscle pain and mechanical hyperesthesia in the orofacial region: a positron emission tomography study. *Pain* 2004; 108: 284-293.

[85] Maihofner C, Forster C, Birklein F, et al. Brain processing during mechanical hyperalgesia in complex regional pain syndrome: a functional MRI study. *Pain* 2005; 114: 93-103.

[86] Maihofner C, Handwerker HO, Neundorfer B, et al. Patterns of cortical reorganization in complex regional pain syndrome. *Neurology* 2003b; 61: 1707-1715.

[87] Gebhart GF. Descending modulation of pain. *Neurosci. Biobehav. Rev.* 2004; 27: 729-737.

[88] Ochoa JL. Comment on: A role for the brainstem in central sensitization in humans. Evidence from functional magnetic resonance imaging. Zambreanu et al. Pain

2005;114:397-407. *Pain* 2005; 117: 236; author reply 236-237.

[89] Schmidt R, Schmelz M, Torebjork HE, et al. Mechano-insensitive nociceptors encode pain evoked by tonic pressure to human skin. *Neuroscience* 2000; 98: 793-800.

[90] Porro CA, Cavazzuti M, Baraldi P, et al. CNS pattern of metabolic activity during tonic pain: evidence for modulation by beta-endorphin. *Eur. J. Neurosci.* 1999; 11: 874-888.

[144] Ushida T, Ikemoto T, Taniguchi S, et al. Virtual pain stimulation of allodynia patients activates cortical representation of pain and emotions: a functional MRI study. *Brain Topogr.* 2005; 18: 27-35.

*Chapter 2*

# COMPARING THE LIMITS OF 3D CONTRAST ENHANCED MRA AT 1.5 AND 3T TO ACHIEVE HIGH SPATIAL RESOLUTION

## *Osama Al-Kwifi* and Graham A. Wright*
Medical Imaging Research, Sunnybrook and
Women's College Health Sciences Centre,
Toronto, Ontario, Canada

## ABSTRACT

*Purpose:* To investigate signal-to-noise ratio (SNR) behavior at 1.5T and 3T using different spatial resolutions with 3D contrast-enhanced magnetic resonance angiography (CE MRA) and to evaluate the role of an 8-channel coil and the parallel imaging technique in trading off SNR and scan time.

*Materials and Methods:* An automatically triggered 3D CE MRA protocol was performed at both field strengths. The intracranial arteries were imaged under four different scenarios, changing the spatial resolution and coil setting.

*Results:* SNR is doubled at 3T compared with 1.5T. At higher spatial resolution (0.5mm isotropic) SNR drops dramatically at both field strengths, as expected. Image quality at 1.5T becomes unacceptable for diagnosis, whereas images at 3T are acceptable. The 8-channel coil is limited in improving SNR values in deep intracranial arteries at 1.5T.

*Conclusion:* Obtaining a 0.5mm isotropic spatial resolution is feasible at 3T, as a result of higher SNR. Parallel imaging may facilitate scan time reduction at 3T for high resolution acquisitions.

**Keywords:** Intracranial arteries, three dimensional imaging, parallel imaging, multi-channel coil, MR angiography.

---

[*] Correspondence: Osama Al-Kwifi, Medical Imaging Research, S611. Sunnybrook and Women's College Health Sciences Centre, 2075 Bayview Avenue, Toronto, Ontario, Canada. M4N 3M5. Tel: (519) 588 -1272; Email: osama@sri.utoronto.ca

## INTRODUCTION

Advantages of high magnetic field strengths, like 3T, have been well established for different applications such as spectroscopy and brain functional imaging, which benefit from the increased SNR. The role of a 3T field strength in magnetic resonance angiography was established recently by evaluating it successfully in different clinical trails [1], demonstrating increased blood vessel contrast and background suppression. The immediate advantage of using 3T over 1.5T is doubling the available SNR [2]. In addition, improved contrast may be achieved as a result of different tissue relaxation times, where longer T1 values at 3T improve background suppression [3]. This could open new avenues for many clinical applications especially MRA, where increasing spatial resolution remains a critical issue for improved assessment of various vascular diseases.

For further SNR improvement, array coils were introduced over standard volume coils. The number and size of coil elements, or channels, will determine local SNR improvements and signal homogeneity. As the size of coil elements is decreased, the SNR close to the coil will improve. Unfortunately, SNR gain varies across the coil zone; an 8-channel head coil increases SNR by at least a factor of 2 near its surface, while yielding a similar SNR in the coil centre [4]. On the other hand, one 16-channel head coil increases SNR by a factor of 6 near the surface and a factor of 2 in the centre [5].

Recently, parallel imaging techniques have been introduced, substantially altering the field of fast MR imaging [6]. These techniques use spatial information contained in the component coils of an array to partially replace spatial encoding which would normally be performed using gradient fields, thereby reducing scan time considerably. Since local SNR is bounded by the square root of imaging time, a SNR penalty arises explicitly. However, by selecting a well-designed array coil and low acceleration factor (rate-2), SNR drop off could be acceptable [7].

In this study, the SNR behavior at 1.5T and 3T using 3D CE MRA with different spatial resolutions has been investigated to illustrate the limitations at both field strengths in producing sufficient SNR for diagnosis in intracranial arteries. In addition, the role of an 8-channel coil, with and without parallel imaging, in improving SNR and/or reducing scan times at high resolutions at both fields is evaluated.

## METHODS AND MATERIALS

### Study Population

After institutional review board approval, an informed written consent was obtained from each patient. 23 patients (mean age 45.6±7.8, 14 male, 9 female) were scheduled to undergo a 3D CE MRA examination. Patients were scanned with a 1.5T or 3T scanner (Signa TwinSpeed, EchoSpeed, and Excite) (GE Medical Systems, Milwaukee, WI) using either standard head coil or 8-channel coil. The head coil and 8-channel array coil have the same configuration at both fields.

## MR Imaging Protocol

An automatically triggered 3D CE pulse sequence [8] with elliptic centric view ordering [9] was positioned in axial view to cover the cerebral arteries including Circle of Wills. The study course was divided into four cases, where the selected parameters of each case were kept as equivalent as possible in both field strengths.

### 1) Low Spatial Resolution with Single Channel Coil

A 3D CE MRA clinical protocol at 1.5T was used with the following parameters: TR/TE 6.6/1.7 ms, flip angle (FA) 30°, field of view (FOV) 22 cm, phase FOV 0.75, matrix 320 × 320, bandwidth 62.5 kHz, slice thickness 0.8 mm, 80 slices, resulting in spatial resolution of 0.68 × 0.68 × 0.8 mm and scan time approximately 2 minutes and 8 seconds. 5 patients underwent this exam. The same parameters were adapted at 3T with TR/TE 6.1/1.6 ms, with all other parameters remaining the same. 4 different patients underwent identical study with these parameters, in addition to 1 patient who was scanned on 1.5T. A theoretical simulation was performed to determine the optimal flip angle that produces maximum signal-to-background contrast. Bloch equations for a series of RF pulses with different flip angles were used in this simulation, considering complete RF spoiling after each RF pulse. In this simulation, T1 background values at both field strengths were taken from the literature [2]. Blood signal intensity at 1.5T and 3T was measured from the source images to calculate T1 of blood. The difference between the background and blood signals (Bloch equation output) was plotted to determine the optimal flip angle that produces maximum contrast. This value was adopted in later studies.

### 2) High Spatial Resolution with Single Channel Coil

To evaluate SNR behavior at higher spatial resolution and determine if better diagnostic images could be generated, 2 patients were scanned at 1.5T using slice thickness 0.5 mm, matrix 416 × 416, and FA 40°, and 3 patients were scanned at 3T with the same parameters, resulting in a spatial resolution of 0.53 × 0.53 × 0.5 mm and scan time of approximately 3 minutes.

### 3) High Spatial Resolution with 8-Channel Coil

To address the SNR reduction at 1.5T and 3T when .5mm isotropic resolution is used, the same parameters from case (2) were used with an 8-channel head coil (MRI Devices Corporation, Waukesha, WI). In this case, 3 patients were imaged at 1.5T and another 2 were imaged at 3T using the identical set up.

### 4) High Spatial Resolution with 8-Channel Coil and Parallel Imaging

The role of parallel imaging in reducing the scan time, which increases dramatically at high resolutions and could affect signal behavior due to contrast dynamics, was evaluated by running the ASSET technique with an ASSET factor of 2 [10] along with the 8-channel head coil. 2 patients were imaged at 1.5T and 2 patients were imaged at 3T with a total scan time approximately 1 minute and 30 seconds.

## 2D Real Time Detection Sequence

The 2D contrast-agent detection sequence is launched and controlled directly from the MR console [8]. It consists of a continuously refreshed 2D transverse section of a gradient-echo imaging sequence. The sequence parameters were as follows: TR/TE 11.5/4.5 ms; FA 30°; FOV 20 cm; matrix 320 × 128; bandwidth 62.5 kHz; and section thickness 10mm. The resultant image update rate was approximately one image every 1.5 seconds. After starting the 2D detection sequence, a region of interest was selected over the basilar and carotid arteries. Inferior and superior saturation bands were sequentially applied and left active during the detection sequence period to cancel signal intensity due to flow-related enhancement within the carotid arteries; thus the signal changes observed will be due primarily to the arrival of gadolinium-enhanced arterial blood.

The signal-intensity-triggering threshold was determined from the mean and standard deviation (SD) for the brightest 20% of the pixels in the region of interest (ROI), calculated over 10 consecutive images. A triggering threshold was set at 6 SDs higher than the mean of the ROI signal intensity over the 10 images. Subsequent intravenous contrast injection of gadolinium-based material resulted in a rapid increase in signal intensity within the regions of interest that exceeded the calculated threshold. Upon triggering, the detection sequence was automatically terminated and the 3D CE sequence was simultaneously started. This auto detection technique was previously presented in detail [8].

## Contrast Agent Injection

A two-cylinder MR imaging compatible injector (Spectris; Medrad, Indianola PA) was loaded with 30 mL of gadolinium-based agent (Omniscan; Nycomed Amersham, Buckinghamshire, England) in one syringe, and with 30 mL of normal saline in a second syringe. With the 3D sequence set up and the detection sequence armed and running as previously described, 30 mL milliliters of the gadolinium-based contrast material was injected at a rate of 3 mL/second. This was immediately followed by a 30 mL bolus of normal saline injected at 3 mL/second. This injection protocol was followed for all patients at both fields.

## Image Evaluation

Source images from each MR exam were transferred to a workstation with 3D capability (Advantage Windows 4.0, GE Medical Systems, Milwaukee WI) for further image analysis. For an objective measure of image quality, SNR value is calculated from the source images of each exam. It has been shown that observers consider image quality to be adequate when SNR is greater than 20 [11]. Measurements were made over the internal carotid artery. Signal intensity (SI) measurement was performed by using an ROI in the centre of the vessel segment. This was the region with the highest and most uniform SI. Noise measurement was defined as the expected noise SD in the ROI (STDnoise). It was calculated from a measurement of signal variance in an ROI in the background in air outside the body, corrected for the effect of the magnitude operator on regions with no signal (Rayleigh

Distribution) [12]. The SNR value is calculated from SI and STD measurements as follows: SNR = SI / STDnoise.

Maximum intensity projection (MIP) images were obtained for each exam for comparison. The evaluation criteria for overall MIP image quality is made by a radiologist and ranked as acceptable/unacceptable. From this evaluation the required SNR level to get acceptable diagnostic images is estimated. All spatial resolutions stated above were attained without interpolation techniques (zero filling).

## RESULTS

SNR measurements for all cases at both field strengths are shown in Table 1. The simulation in case (1) revealed that a 40° flip angle optimizes contrast between blood and background at both fields. In case (1), SNR is doubled at 3T compared with 1.5T in agreement with [1]; image quality at both 1.5T and 3T was considered acceptable for clinical diagnosis. However, background suppression and vessel conspicuity were higher at 3T, as shown in Figure 1.

**Table 1. SNR measurement for different spatial resolutions and head coils**

| Study Case | Spatial resolution / Scan time / head coil | SNR (1.5T) | SNR (3T) |
|---|---|---|---|
| First | 68x.68x.8 / 2:08 minutes / single channel | 23.3 ± 2.4 | 39.6 ± 4.2 |
| Second | 53x.53x.5 / 3:00 minutes / single channel | 11.4 ± 1.3 | 24.6 ± 3.1 |
| Third | 53x.53x.5 / 3:00 minutes / 8-channel | 10.8 ± 0.9 | 25.1 ± 2.4 |
| Fourth | 53x.53x.5 / 1:30 minutes / 8-channel + ASSET | 8 ± 0.7 | 20.6 ± 2.3 |

All measurements were made over the internal carotid artery.

In case (2) with 0.5mm isotropic resolution, SNR drops dramatically for both fields resulting in images evaluated as unacceptable at 1.5T, whereas images quality at 3T remain acceptable for diagnosis, as shown in Figure 2.

In case (3), using an 8-channel coil at 1.5T and 3T improves SNR, particularly near the head surface [4], Figure 3. However, at the centre of the head, where our measurements were made, the SNR remains comparable to the standard head coil. This led to poor images at 1.5T, whereas images at 3T were acceptable for diagnosis.

In case (4), adding parallel imaging to the 8-channel coil reduces the scan time to half at the cost of reduced SNR, especially in the central region of the coil, as illustrated in Table 1. In this scenario, image quality was considered unacceptable at 1.5T and acceptable at 3T, as shown in Figure 4. Despite the drop-off in SNR associated with parallel imaging, vessel edges are still well defined at 3T.

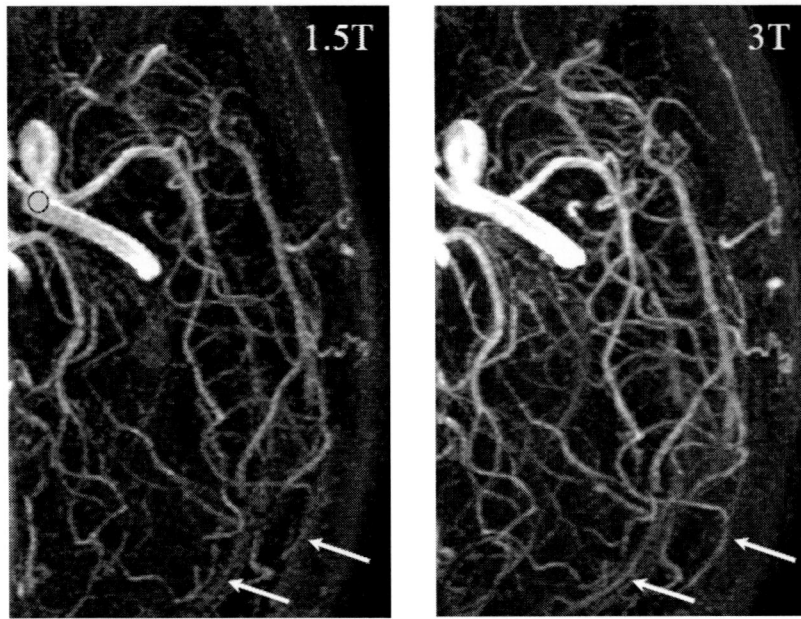

Figure 1. Intracranial MIP images for the same patient at spatial resolution of 0.68 × 0.68 × 0.8 mm at both field strengths, case (1). Image quality at both fields is good and has acceptable diagnostic value. Superior background suppression and better vessel delineation are noticed at 3T, especially for the distal vessels. Gray circle indicates the ROI in the source image, where signal intensity was measured.

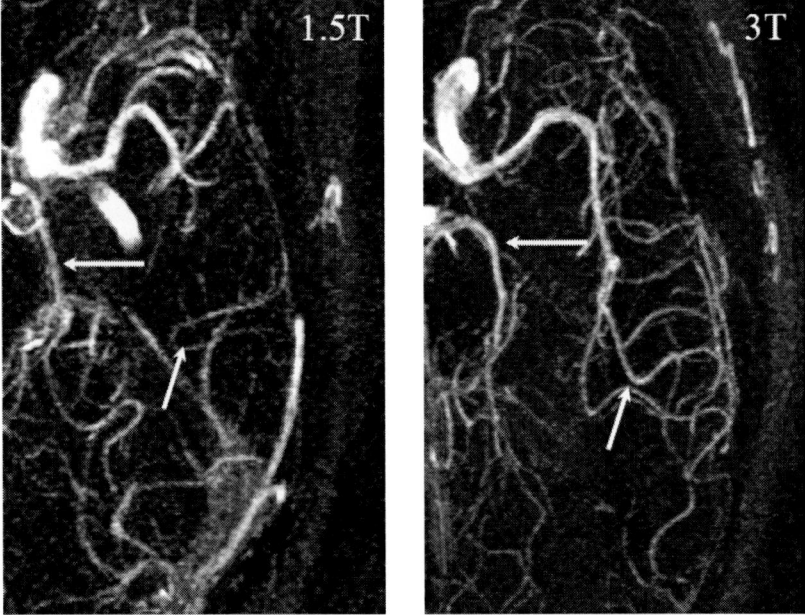

Figure 2. Intracranial MIP images for two patients at spatial resolution of 0.53 × 0.53 × 0.5 mm, case (2). It is clear that the SNR reduction at higher resolution affects vessel conspicuity at 1.5T, producing low quality images, whereas higher SNR at 3T sufficiently compensates for this reduction, generating good vessel depiction and acceptable images.

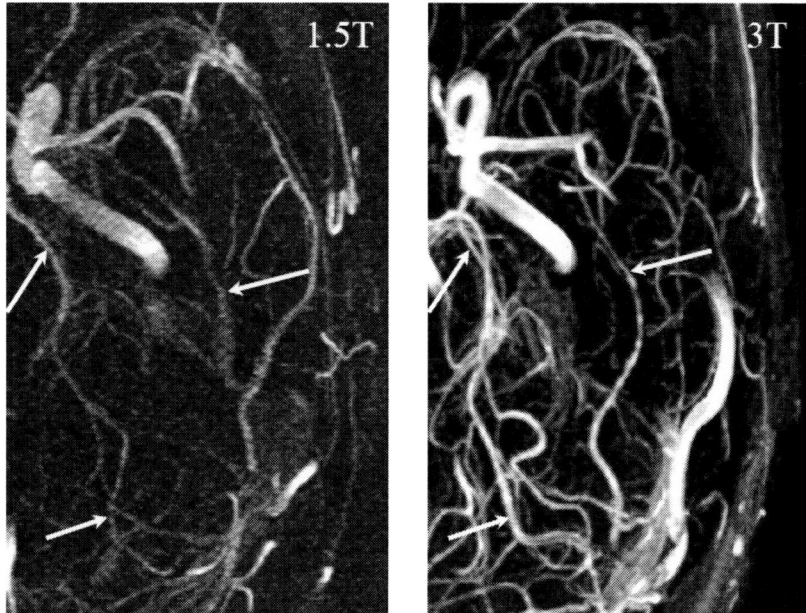

Figure 3. Using an 8-channel coil at spatial resolution of 0.53 × 0.53 × 0.5 mm, case (3). The SNR increases near the head surface in the regions close to the head coil, at both field strengths, whereas at the central region of the head the SNR value does not change substantially compared to case (2). This keeps the overall image quality at 1.5T unacceptable, while 3T image quality remains acceptable.

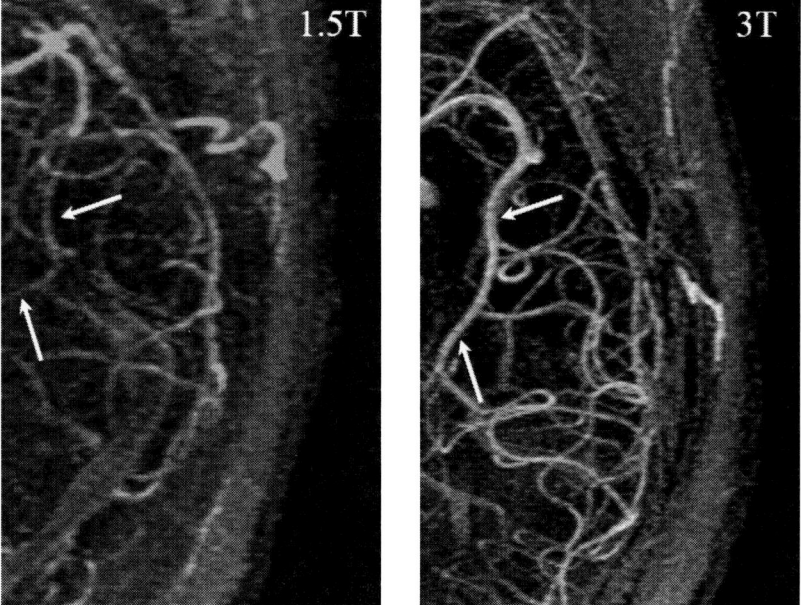

Figure 4. Deploying a parallel imaging technique with an 8-channel coil at spatial resolution of 0.53 × 0.53 × 0.5 mm, case (**4**). It is evident that SNR drops dramatically as the scan time decreased to half. Vessels close to the head surface become poorly visualized at 1.5T, leading to unacceptable image quality. At 3T, SNR reduction associated with parallel imaging is still tolerable, where the major superficial arteries can be identified easily.

## DISCUSSION

The immediate gain of deploying 3T over 1.5T is doubling the SNR, leading to increased conspicuity of the fine vasculature and reduction in the graininess of the image, as confirmed in preliminary work [1,2]. In our study, this has been demonstrated clearly at a spatial resolution of 0.68 × 0.68 × 0.8 mm. At higher spatial resolution (e.g. 0.5 mm isotropic), the resulting voxel size would lead to a grainy image at 1.5T, as shown in Figure 2, where the vessel edges are distorted due to the reduced signal from the smaller voxel size. At this spatial resolution, SNR drops almost to half at both fields. On the same figure, vessel edges from the 3T image are well defined without noticeable artifacts. This is due mainly to the increased SNR at 3T. It is apparent that 1.5T is unable to produce good quality images at 0.5 mm isotropic spatial resolution under these conditions.

Utilizing the 8-channel coil with 1.5T revealed some signal improvement in the vessels near the head surface; however no significant change in signal is observed in the central region of the head, as shown in Table 1. At 3T, the 8-channel coil has doubled the SNR at the surface vessels and in the central region the SNR remains comparable to the standard head coil, producing high resolution images with high quality. A further study using an improved coil design, such as 16-channel coil, could enhance the overall SNR value at 1.5T images to generate higher resolution, where the signal in the central region of the head coil could increase by a factor of 2 and near the surface by a factor of 6 [5]. This type of coil was not available in our institution for comparison studies.

In case (4), the use of parallel imaging technique reduces the imaging time by half, but the reduction in SNR is not as great as the decrease in scan time would predict. This suggests that higher contrast concentration at earlier stages of the scan may mitigate some of the predicted SNR loss [10]. In this case, the resulting image quality is acceptable indicating that an SNR of about 20 is desirable for diagnostic images [11].

Since sensitivity falls off faster for a given coil set, the acceleration factor is typically limited to four or less at 1.5T due to the limited SNR. Whereas higher acceleration factors are possible at 3T due to greater intensive SNR and reduced degradation associated with processing for parallel imaging. Specifically, higher field strength improves geometry factors in parallel imaging [13]. Thus 3T, would facilitate new applications such as scanning large volumes with higher spatial resolutions in a given scan time.

Diseases where diagnosis could benefit from improvement in spatial resolution include aneurysms and atherosclerotic disease. Improved diagnostic confidence in MRA could preclude X-ray angiography studies and allow more confident screening of patients with congenital (or other) predispositions to forming aneurysms.

In conclusion, higher spatial resolutions are feasible at 3T, due to the increased inherent SNR. 8-channel coils are limited in improving SNR values in deep intracranial arteries at high spatial resolution for 1.5T. Given these trade-offs, 3T may be necessary to realize diagnostically useful images at approximately 0.5 mm isotropic resolution in the head with CE MRA. Parallel imaging may facilitate scan-time reduction at 3T for high resolution acquisitions.

## REFERENCES

[1] Bernstein MA, Huston J, Lin C, et al. High-resolution intracranial and cervical MRA at 3.0T: technical consideration and initial experience. *Magn. Reson. Med.* 2001:46:955-962.

[2] Al-Kwifi O, Emery DJ, Wilman AH. Vessel contrast at three Tesla in time-of-flight magnetic resonance angiography of the intracranial and carotid arteries. *Magn. Reson. Imaging* 2002:20:181-187.

[3] Wansapura J, Holland S, Dunn R, et al. NMR relaxation times at 3 Tesla. *J. Magn. Reson. Imaging* 1999:9:531-538.

[4] Gizewski ER, Madewald S, Wanke I, et al. Is more better? Comparison of 4 and 8 channel head coils using standard T2-sequence and iPAT. In: *Proceedings of the 11th Annual Meeting of the International Society of Magnetic Resonance in Medicine*, Toronto, Canada, 2003. Abstract 2639.

[5] Porter JR, Wright SM, Reykowski A. A 16-element phased-array head coil. *Magn. Reson. Med.* 1998:40:272-281.

[6] Pruessmann KP, Weiger M, Scheidegger MB, et al. SENSE: sensitivity encoding for fast MRI. *Magn. Reson. Med.* 1999:42:952-962.

[7] de Zwart JA, Ledden PJ, Gelderen PV, et al. Signal-to-noise ratio and parallel imaging performance of a 16-channel receive-only brain coil array at 3.0 Tesla. *Magn. Reson. Med.* 2004:51:22-26.

[8] Farb RI, McGregor C, Kim JK, et al. Intracranial arteriovenous malformations: real-time auto-triggered elliptic centric-ordered 3D gadolinium-enhanced MR angiography--initial assessment. *Radiology* 2001:20:244-51.

[9] Wilman AH, SJ Riederer. Performance of an elliptical centric view order for signal enhancement and motion artifact suppression in breath-hold three-dimensional gradient echo imaging. *Magn. Reson. Med.* 1997:38:793-802.

[10] Chen Q, Quijano CV, Mai VM, et al. On improving temporal and spatial resolution in 3D contrast-enhanced body MRA with parallel imaging. In: *Proceedings of the 11th Annual Meeting of the International Society of Magnetic Resonance in Medicine*, Toronto, Canada, 2003. Abstract 1343.

[11] Owen RS, Wehrli FW. Predictability of SNR and reader preference in clinical MR imaging. *Magn. Reson. Imaging* 1990:8:737-745.

[12] Henkelman RM. Measurement of signal intensities in the presence of noise in MR images. *Med. Phys.* 1985:12:232-233.

[13] Wiesinger F, Boesiger P, Pruessmann KP. Electrodynamics and ultimate SNR in parallel MR imaging. *Magn. Reson. Med.* 2004:52:376-390.

*Chapter 3*

# RAPID ESTIMATION OF MOTION BLUR FOR REAL-TIME OPTIMIZATION OF HIGH RESOLUTION IMAGING[#]

### *Osama Al-Kwifi*[*] *and Graham A. Wright*

Imaging Research Program, Sunnybrook and
Women's College Health Science Centre,
Toronto, Ontario, Canada

## ABSTRACT

Due to substantial carotid and coronary motion variation among individuals it is essential that optimal imaging parameters be determined by characterizing this motion on a subject-by-subject basis. In this work, new methods to compute the effects of motion on different k-space trajectories are introduced. After measuring in-vivo motion a simulation was performed to calculate motion blurring effects using different methods. Final results from each method were evaluated for their accuracy and speed in estimating the motion effects. It has been demonstrated that, for small motions corresponding to reasonable target resolutions, simple approximation methods yield reasonable estimates of motion blur significantly faster than full computations of the point spread function, and could be used to facilitate real-time scan parameter optimization to achieve high resolution imaging.

**Keywords:** Point spread function, high resolution imaging, real time optimization.

---

[#] Some of the results in this paper have been presented in the Angio-Club Dublin, Ireland 2003.
[*] Address reprint requests to O.A.K. Medical Imaging Research, Sunnybrook and Women's College Health Sciences Centre. 2075 Bayview Avenue. Toronto, Ontario, M4N 3M5. Canada. Fax: (416) 480-5714; Tel: (416) 480-6100 (x1018); Email: osama@sten.sunnybrook.utoronto.ca

## INTRODUCTION

Previous studies have shown the significant variations in carotid and coronary artery motion among individuals [1,2]. Therefore, to achieve high resolution images (<1mm) it is important that scan parameters are optimized on a subject-by-subject basis [1,2]. These parameters could include: initial spatial resolution, acquisition window during the heart cycle, trigger delay time, and number of slices acquired per heart cycle. Characterizing the effects of these parameters given the motion characteristics of a specific subject can be made by calculating the full width at half maximum (FWHM) of the point spread function (PSF), which represents a conventional method to measure any image blurring due to motion effects [3]. This optimization requires numerous calculations in order to search the full parameter space and its interaction with other parameters. In this work, to make this optimization more tractable for real-time characterization, we are evaluating two different approximation methods for their accuracy and speed in estimating the effects of vessel motion. Specifically, we quantified the width of the PSF and its relative positional shift from slice to slice. These specific measurements are indicative of blur and distortion problems that limit the ability of magnetic resonance (MR) imaging to make accurate estimate of different carotid and coronary diseases. In this study, we use coronary artery motion as an extreme example of large motion to test the robustness of the new approximation methods. However, for carotid artery motion, which is a fraction of coronary motion, the new methods would operate with higher accuracy due to small motion variations.

## METHODS AND MATERIALS

A 1.5T MR system (GE Signa CV/i, Milwaukee, WI) equipped with a 5" surface coil was used to image the right coronary artery (RCA) from eight healthy volunteers. An interactive real-time tool [4] was used to image the proximal and distal locations of the RCA over a 20s breath-hold. The coronary artery motion for both segments was then measured off-line [5]. To calculate the expected motion blur and coronary position offset resulting from data acquisition at different portions of the heart cycle for a given protocol, a computer simulation using three methods was performed. The first method represents the gold standard whereas the second and third methods are the approximation methods. This simulation was performed assuming a 2D spiral scan [6] with multiple interleaves of 4096 samples with repetition time (TR) 40ms, and field of view (FOV) 20cm. In each method, the set of coronary motion displacements are the motions at the time of each spiral interleaf acquisition assuming a given number of interleaves. These sets of displacements were used to determine the ability of each method in measuring image blur and slice offset.

(1) Conventional Method or Method (1): where the simulated k-space acquisitions of a point object for a given k-space filling trajectory were modulated by a phase shift according to the motion displacements. A simulated image was then reconstructed and the FWHM of the resulting PSF and the shift of the PSF peak from the origin determined.

(2) PSF Convolution Method or Method (2): where an idealized 2D PSF image is generated using a first-order Bessel function ($J_0(x)$) such that its FWHM is equal to the nominal spatial resolution (i.e. with no motion present), as shown in Figure 1A. The initial position of the PSF was located in the image center. This image was convolved with the 2D motion displacements to create a set of images, one for each motion displacement. This set of images was then summed together and the FWHM and the mean shift from the origin determined from the final result.

(3) Motion Distribution Method or Method (3): where the mean and standard deviation of the motion displacements were calculated and related to the PSF offset and FWHM respectively of method (1).

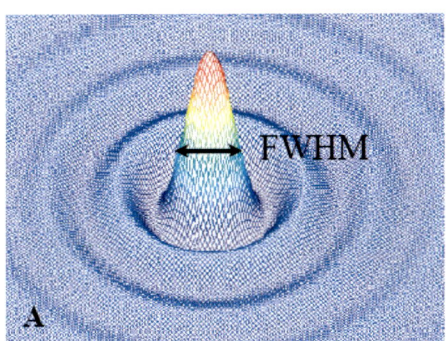 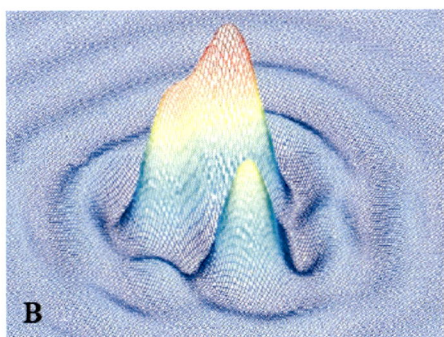

Figure 1. (A) An example of the nominal 2D point spread function whose FWHM is equal to the initial spatial resolution, i.e. in the absence of motion. (B) The summation of multiple nominal 2D PSF that are each shifted according to a set of motion displacements corresponding to the motion at each data acquisition interval. When the motion displacements are large relative to the nominal FWHM the PSF begins to contain multiple peaks and estimating motion blur becomes difficult.

To compare methods (1) and (2), single slice real-time scans with different numbers of interleaves (spatial resolutions) within a single heart cycle were simulated. In this case, increasing the spatial resolution means acquiring more spiral interleaves during a single heart cycle, which cause data acquisition to occur over more of the heart cycle facing larger motion variations. To evaluate the accuracy of method (2) in measuring FWHM, the error in estimating the motion blur using method (2) was calculated as ($[FWHM_{(method.1)}-FWHM_{(method.2)}]/FWHM_{(method.1)}$) and plotted as a function of the ratio between motion amplitude relative to spatial resolution.

Certain clinical applications require larger coverage to define coronary artery anatomy precisely, where multiple slices could be acquired [6]. However, selecting the optimal number of slices that fit in one heart cycle could minimize slice-to-slice misregistration due to motion variation across heart cycles. For this optimization, a simulation of 12 slices breath-hold exam, where each slice consists of 14 spiral interleaves (0.8mm spatial resolution) over 14-heart cycles, was made to compare methods (1) and (3). For each interleaf, data acquisition occurs at the same position (trigger delay time) in each of the 14-heart cycles.

Finally, an optimization approach was used to quickly determine the minimum achievable resolution of the simulated image acquisition given the set of measured motion displacements. Using a gradient descent approach [7] the full parameter space of cardiac timing parameters (acquisition window duration and trigger delay time) was investigated with an additional constraint of the total number of heart cycles (total scan time) for the image acquisition. Optimizations were performed for acquisitions covering a single heart cycle (simulating a real-time scan) or multiple heart cycles (simulating a breath-hold scan). The processing time for typical optimizations was estimated based on the total computational complexity and typical measured computation time for each step in the analysis.

This simulation is specific to the multi-slice 2D imaging approach [6]; however, by adding appropriate modifications it could also be applied to other types of acquisition strategies, like 3D imaging approach.

## RESULTS AND DISCUSSION

Figure 2 shows that estimating motion blur in method (2) gives comparable results to method (1) as long as the motion amplitude is equal or less than the initial spatial resolution. Within this range motion blur calculated using method (2) is strongly correlated to the FWHM from method (1) (R=0.98). Beyond this point the motion displacements become large relative to the initial width of the PSF such that when one applies method (2) the final PSF begins to contain multiple, distinct peaks as shown in Figure 1B. In these circumstances quantifying the FWHM in a meaningful way can be difficult. However, in such situations the blur is at a level which is unacceptable for imaging in the first place. Within the same previous range, mean slice offset as measured with method (2) is strongly correlated to the mean offset measured using method (1) (R=0.97).

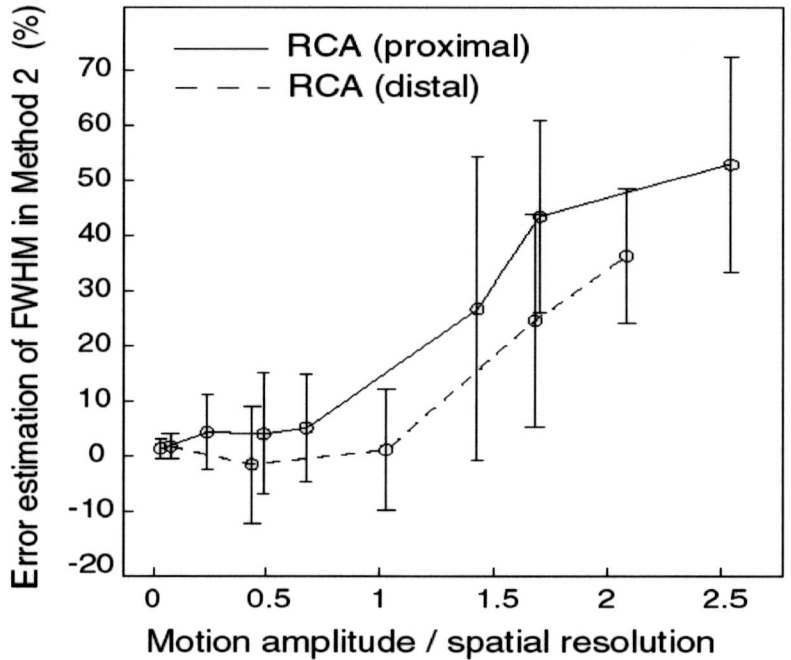

Figure 2. Error in estimating motion-induced blur using Method (2). When the motion amplitude exceeds the spatial resolution, errors in measuring FWHM become excessive.

Figure 3 shows PSF slice offsets (PSF peak values) for 12 different slices collected over 14 heart cycles using method (1) and (3). The slice offset position calculated using method (3) is strongly correlated to the actual offset position calculated using method (1) (R=0.97). The mean offset results from collecting data at different portions of the heart cycle. It is obvious that method (3), is capable of estimating much larger offset shifts than method (2). It may therefore be appropriate in determining the optimal number of slices to be collected within a heart cycle with minimal slice-to-slice distortion. When comparing the FWHM of method (1) and the standard deviation of method (3) there is no correlation between these measures (R=0.59) indicating that method (3) does not provide a reasonable estimate of the expected amount of blur due to motion.

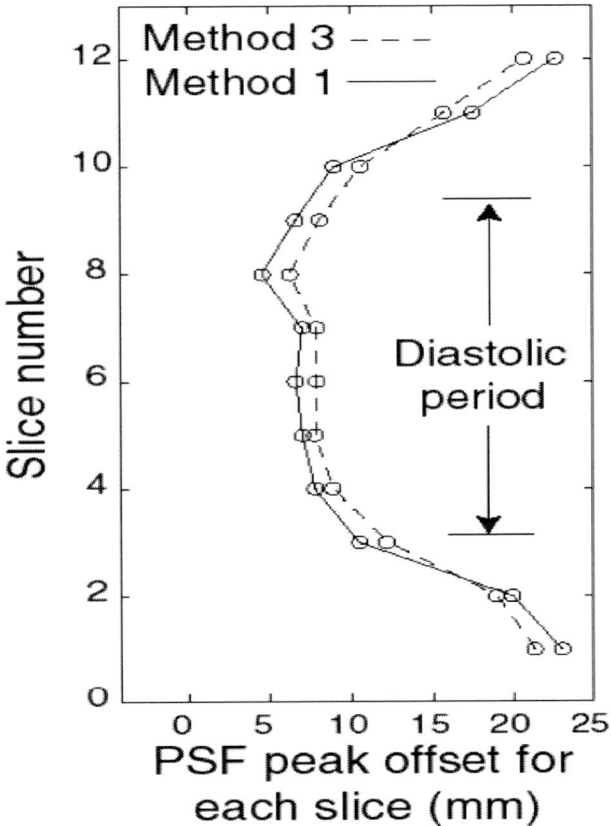

Figure 3. PSF peak offset for 12 different slices acquired at different phases in the heart cycle during a breath hold scan. Method (3) gives accurate estimates of the mean slice displacement compared to method (1), which can determine the optimal number of slices to collect per heart cycle.

The optimization processing time of motion blur using method (2) when applied to the single and multiple heart cycle cases required 800ms and 2.5s respectively on a PIII 700MHz machine. The general order of calculations was used to compare the computational costs of the three different methods. Each FWHM estimate using method (2) takes about 20-25ms ($N^2*nl$, N: matrix size, nl: number of interleaf). Method (1) requires applying: gridding ($nl*res*9$, res: number of samples), phase-shift ($N^2*nl$), and Fourier transform ($(N*4)^2 *\log(N*4)$ ). In the Fourier transform step, zero filling by a factor of four was used to allow the measurement of the FWHM to an appropriate degree of accuracy. The expected improvement in computation time is ~10 times for method (2) over method (1). The computation time in Method (2) can be further reduced by limiting the extent of the matrices over which the PSF is evaluated (which will depend on the range of motion displacements). For most reasonable cases an additional improvement in computation time by a factor of 4 can be achieved by cropping the PSF matrices by a factor of 2 in each direction. Finally, the computation time of Method (3) (2*nl) is 10,000 times faster than method (1).

## CONCLUSION

Conventional Method (method 1) is the gold standard in calculating motion blur and slice offset, but using it for pre-scan optimization would take additional time. However, the introduced approximation methods are much faster and able to do the same evaluation within limits. Both methods could be used together to determine the optimal scan parameters, where PSF Convolution Method (method 2) can verify the blurring for each slice within the heart cycle and Motion Distribution Method (method 3) sets the total number of slices that could fit for that particular subject.

## REFERENCES

[1] O. Al-Kwifi, Jae Kim, Jeff Stainsby, et al. Pulsatile motion effects on 3D magnetic resonance angiography: implications for evaluating carotid artery stenoses. *Magnetic Resonance in Medicine* 2004; 52: 605-611.

[2] O. Al-Kwifi, J. Stainsby, F. Warren, et al. Characterizing coronary motion and its effect on MR coronary angiography--initial experience. *Journal of Magnetic Resonance Imaging* 2006; 24: 857-869.

[3] E.M. Haacke, R.W. Brown, M.R. Thompson, R. Venkatesan, Magnetic resonance imaging: physical principles and sequence design. New York: John Willey and Sons, 1999, pp. 269-271.

[4] A.B. Kerr, J.M. Pauly, B.S. Hu, K.C. Li, et al. Real-time interactive MRI on a conventional scanner," *Magnetic Resonance in Medicine* 1997; 38: 355-367.

[5] M.S. Sussman, G.A. Wright, "Factors affecting the correlation coefficient template matching algorithm with application to real-time 2D coronary artery MR imaging," *IEEE Medical Imaging Transaction* 2003; 22: 206-216.

[6] C.H. Meyer, B.S. Hu, D.G. Nishimura, A. Macovski, Fast spiral coronary artery imaging. *MagneticResonance in Medicine* 1992; 28: 202-213.

[7] K. Atkinson. An Introduction to Numerical Analysis. 2nd ed., New York: John Wiley and Sons, 1989.

## Chapter 4

# ELECTRIC SOURCE IMAGING AS A TECHNIQUE TO QUANTITATIVELY DETERMINE ONSET AND SPREAD OF INTERICTAL AND ICTAL EPILEPTIFORM ACTIVITY

### Göran Lantz[a,b], Pål G. Larsson[b] and Christoph M. Michel[a]

[a] Functional Brain Mapping Laboratory, Neurology Clinic,
University Hospital, Geneva Switzerland
[b] The National Centre for Epilepsy, Sandvika Norway

### ABSTRACT

In this paper the recent development in Electric Source Imaging (ESI) in epilepsy investigations will be discussed. ESI has made tremendous progress in the past years and is now available as an additional tool in the presurgical workup of patients with pharmaco resistant partial epilepsy. Many different source models have been proposed, each with their advantages and disadvantages, and it is therefore advisable to perform the source reconstructions not with one single model but rather with a number of these different techniques. It is also important that the reconstructions are performed using a realistic head model, preferably the MRI of the individual patient, and that a sufficient number of recording electrodes (at least 60-70) are used. The results can also be improved by using statistical SPM like techniques to determine areas of significant activation. In the paper these different methodological issues are addressed. Some of the literature on interictal and ictal source propagation is reviewed, and it is discussed how ESI can be used to characterize the focus of interictal and ictal epileptiform activity, in terms of focus localization but also in terms of visualizing these complex propagation patterns. Accurately determining the anatomincal origin of interictal and ictal epileptiform activity from EEG recordings is of crucial importance in presurgical epilepsy investigations, and the possibility to characterize interictal and ictal propagation patterns also gives important additional information for understanding the epileptic process in the individual patient, for instance in relation to the ictal semiology or to the results of other neuroimaging techniques.

## A. INTRODUCTION

Electroencephalography (EEG) is an important technique for studying the temporal dynamics of the neuronal circuits in the human brain. Especially in patients with epilepsy it provides an excellent tool to visualize the spatial and temporal variations of the abnormal electrical activity that leads to the clinical manifestations of the epileptic disease. However, most EEG applications do not fully capitalize on all of the data's available information, particularly when it comes to determining the localization of the different active sources in the brain.

Especially in patients with pharmacoresistant partial epilepsy, in whom surgical resection of the epileptic focus is often considered, it is of major importance to make a correct and anatomically precise identification of the epileptogenic region. In these presurgical investigations, which are normally performed at highly specialized epilepsy centers, multichannel EEG synchronized with Video-monitoring is performed continuously for several days in order to assess the EEG patterns that are typical for the patient's seizures. These patterns are usually judged by visual inspection both of the EEG traces during interictal discharges and before, during and after seizures. Together with neurological and neuropsychological exams, high-resolution MRI, PET, interictal and ictal SPECT and MRI-based volumetry and spectroscopy, the results of the EEG examinations are then used to delineate the epileptogenic focus with the highest possible accuracy, and to constitute the foundation of the decision about which areas of the brain should be resected. In cases where the results of these exams are unclear or non-concordant, additional recordings with electrodes implanted in the brain may be necessary [1].

Although visual EEG analysis may be quite accurate in the hands of an experienced epileptologist, the limitations are also well known [2], and it has been suggested that the basic trace analysis of the EEG only capitalizes from a fraction of all the information that is actually present in the signals. [3]. Since nowadays virtually all clinical EEG-labs have moved from paper-EEG to digitally recorded EEG, advanced analysis of the signals with modern signal processing tools have become possible. Consequently, in recent years, techniques to quantitatively localize the source of interictal and ictal epileptiform activity have been developed, and in order to emphasize the analogy between electric source locations and other imaging techniques, the term Electric or EEG Source Imaging (ESI) has been introduced [4].

In this review the recent development in ESI of epileptic data will be addressed. We here concentrate on EEG recordings, though most of the aspects discussed similarly concern MEG. Similarities and differences between EEG and MEG have been discussed elsewhere [5; 6; 7; 8; 9].

## B. METHODOLOGICAL CONSIDERATIONS

There are a number of different factors that influence the accuracy of the source localizations in Electrical Source Imaging (ESI). Some of these factors that are going to be addressed in the following are: (i) Which source model should be used? (ii) How can the ESI results be integrated in the patient's own anatomy? (iii) How many recording electrodes are needed? (iv) How can the significance of a certain activation pattern be determined?

## 1. Which Source Model Should Be Used?

The main task in ESI is the solution of the so called inverse problem. This problem consist in estimating the number and location of electric sources in the brain based on the voltage distribution recorded on the surface of the skull. The inverse problem is ill posed, which means that a certain surface voltage distribution may be explained by several different combinations of source configurations in the brain. For this reason it is necessary to impose some restrictions on the solution in order to identify the most likely of the many different possible source configurations. These restrictions may be mathematical (for instance the solution with the smoothest distribution in space is selected) or anatomical/physiological (sources are only allowed in grey matter). There are two classes of inverse solutions, linear or overdetermined models (such as the dipole model [10]) which postulates one or a few spatially limited sources in the brain, and non-linear or distributed source models [11, 12; 13] with which no such assumptions have to be made. Different studies applying these different techniques to epileptic data will be presented below. Each of the different source models have their advantages and limitations which have been addressed in other review articles [14, 15, 16, 17; 18]. The question about which model to use was raised in a special issue of the journal "Clinical Neurophysiology" dedicated to EEG source modeling [19], where the editor concluded that "All methods, even if they work well under most conditions, may fail under others", and that "Solving the inverse problem using a number of different techniques is thus more prudent and more likely to lead to reasonable results".

## 2. How Can the ESI Results Be Integrated in the Patient's Own Anatomy?

A prerequisite for obtaining correct source localizations is proper selection of the head model for the inverse solution calculation. The head model determines the relationship between the sources located at given brain sites and the measurements on the scalp. It includes the electromagnetic (permeability's and conductivities) and geometrical (shape) properties of the volume within which the inverse solutions are calculated. In previous source modeling studies the source reconstructions have often been performed in a standard spherical head model with uniform conductivities for brain skull and scalp. This is an approach which in most situations will lead to a rather coarse approximation of the true source location, although the situation to some extent can be improved through incorporations of different conductivity parameters in multi-shell spherical head models [20, 21, 22, 23] and by considering local anisotropies [24, 25].

More recently, techniques have been developed that permit definition of the source coordinates in terms of Brodmann areas or in Talairach coordinates. Apart from increasing the accuracy of the ESI results this approach also makes it possible to draw conclusions about the active anatomical/functional structures. To be able to define the inverse solution results within the structural MRI, and thus to determine which anatomical areas are active at a certain timepoint a so called EEG-MRI coregistration between the EEG space and the MRI space is performed. In this step electrode positions are matched to the scalp surface defined by the MRI using some transformation operations (rotation and translation). These parameters are usually obtained by defining some "common" landmarks both during the EEG and the MRI acquisition [10, 26, 27, 28, 29, 30, 31].

The mostly used techniques for realistic head modeling are the BEM (boundary element method, [32]) and the FEM (finite element method [33, 34, 35]). In practical terms the difference between the two techniques is that, in contrast to the BEM model, the FEM model is able to take anisotropies in the brain into account. A major disadvantage with these techniques is that the solutions have to be calculated numerically, which requires a considerable computational load. As an alternative to the above described head models, a spherical head model with anatomical constrains (SMAC) has been proposed [36]. With this approach a best-fitting sphere is calculated for the individual head surface derived from the segmented MRI, and in this way the inverse problem can be calculated analytically (using a spherical model), but still taking into account the anatomy of the individual patient derived from the slightly deformed MRI. The major advantage as compared to the traditional spherical head is the possibility to directly exclude areas that are not expected to contribute to the solution, such as white matter or lesions.

One example of ESI performed in the patients own MRI is shown in figure 1.

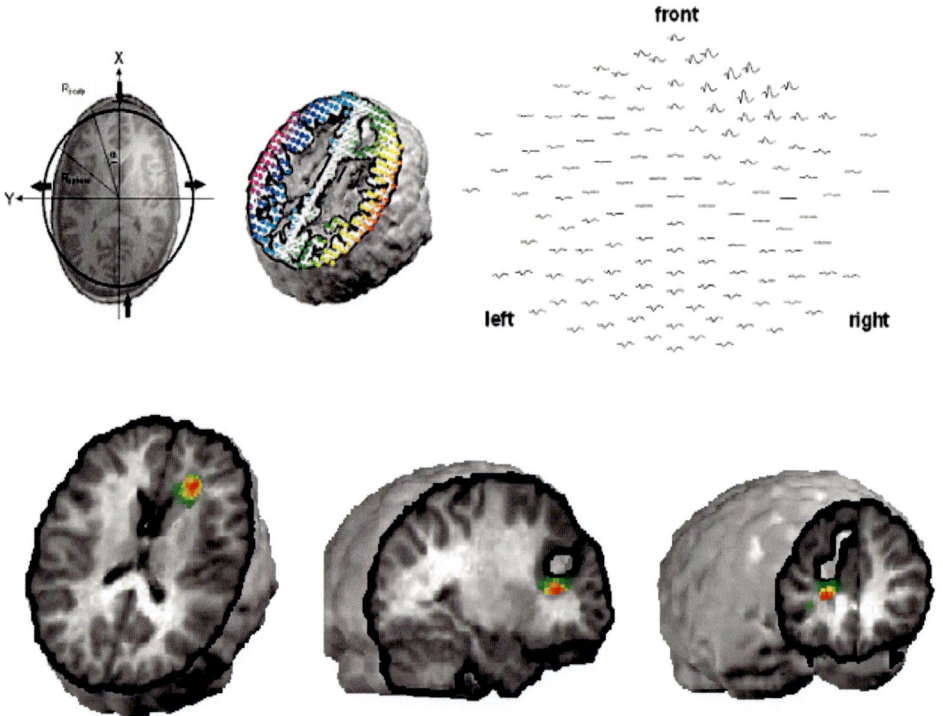

Figure 1. Localization of the source of an interictal spike in the patients own anatomy. (A) The SMAC technique is used to transform the patients MRI into a sphere. (B) The grey matter is identified and the solution points (i.e. the points in the brain where the the soulution is allowed to localize) are identified. (C) EEG of an epileptic spike with a right frontal distribution. (D) The ESI localizes the source close to an anatomical lesion in the right frontal lobe.

A less labor intensive alternative which has been applied in many studies is to use one single template MRI (such as the MNI brain from the Montreal Neurological Institute) and assume that the electrodes are placed in accordance with a standard electrode coordinate system. Although more accurate than a spherical head model, this approach still ignores individual differences in head size and electrode positions, leading to a limited accuracy of the source locations.

A pitfall with the superposition of EEG on the patients MRI is that an erroneous impression of high anatomical precision may be given when spatially limited sources are displayed on the MRI, especially if the reconstruction is performed in a spherical head model without any anatomical restrictions. A dipolar source only represents the centre of gravity of the true intracranial voltage distribution, and also with distributed source models the true extent of the activated regions is impossible to determine. Especially for epileptiform activity, which may have a very irregular distribution within the brain volume, an over-detailed interpretation of the source localizations should be avoided [19].

## 3. How Many Recording Electrodes Are Needed?

Another requirement for correct ESI results is a sufficient number of electrodes to adequately sample the surface electric field. From theoretical studies it has been concluded that the required number of electrodes would be over 100 [37; 38; 39].

So far relatively few studies using high density EEG recordings in epilepsy have been presented, and in most of the existing studies only 64 channels have been used [40; 41; 42]. In a recent investigation [43] 14 epileptic patients were investigated during the course of a presurgical evaluation. Interictal epileptiform activity was recorded with 125 electrodes, and each single epileptiform potential was downsampled to 63 and 31 electrodes. Subsequently a distributed source model (EPIFOCUS)[44] was used to reconstruct the sources of the epileptiform activity with the 3 different electrode configurations. The localization accuracy with the 3 electrode setups was assessed by determining the distance from inverse solution maximum of each single spike to the epileptogenic lesion. In 9/14 patients the distance from the EEG source to the lesion was significantly smaller with 63 than with 31 electrodes, and this ratio was marginally increased by increasing the number of electrodes to 125. Simulations confirmed the relation between number of electrodes and localization accuracy.

From the latter study it was concluded that at least around 60 electrodes would be necessary for epileptic focus localizations, which is more than the around 30 electrodes nomally used in clinical routine investigations.. This has previously been difficult to achieve in normal clinical practice because of the large amount of work to mount such a number of electrodes. Recently, however, systems have become commercially available, which allow the application of 125-256 EEG electrodes within less than 10 minutes [4; 45].

An example of erroneous source localizations due to insufficient electrode coverage is given in figure 2.

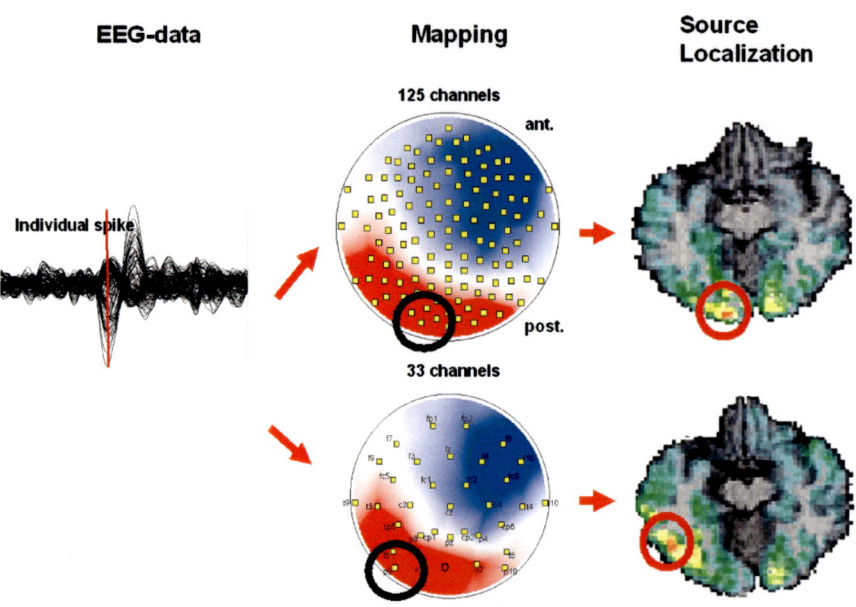

Figure 2. An example illustrating the problem with undersampling of the electric field. The patients has a malformation in the left parieto-occipital cortex. To the left an individual spike recorded with 125 electrodes. Right above: ESI with 125 electrodes correctly localizes the source to the left occipital region. Right below: ESI after symmetrical downsampling to 33 electrodes (standard 1020 montage + some additional electrodes) erroneously localizes the source to the posterior temporal area due to insufficient coverage of basal posterior areas. Please note that the standard occipital electrodes O1 and O2 are used also in the 33 electrode montage.

## 4. How Can the Significance of a Certain Activation Pattern Be Assessed?

With all existing techniques for source reconstruction it is inherently impossible to determine the true spatial extent of an activation pattern. However, if distributed inverse solutions are used for the source reconstructions, statistical analysis similar to that proposed for other tomographic functional imaging techniques can give some indication about the significance of a certain area of activation. In fMRI studies the so-called statistical parametric mapping (SPM) is commonly used. The individual fMRI images are spatially normalized to a standard brain. Then a t-test between two conditions is performed for each voxel or region of interest, in order to to determine areas of significant difference between the two conditions. A similar statistical procedure can be applied to the tomographic maps obtained from the distributed inverse solutions. In [46] the source estimation was performed on the individual BEM-modelled MRI. In other studies, however, a simpler approach has been used by normalizing the maps to a standard head shape configuration using spherical spline interpolation, solving the inverse solution for this general head model using the same solution space for all subjects, and then performing t-test statistics for each voxel of the solution space [47; 48; 49; 50]. Recently such a statistical method has been proposed for interictal epileptic activity by statistically comparing the inverse solutions during the epileptiform discharge with a baseline period [51; 52].

## C. CHARACTERIZATION OF THE INTERICTAL EPILEPTIC FOCUS

### Source Localization of Interictal Epileptiform Activity

In order to solve the inverse problem for epileptic data different techniques have been applied (for reviews see [17; 53; 54; 55]. In earlier studies, most investigators have used the concept of approximating the activity of the epileptic focus, recorded with EEG or MEG, with one or a limited number of dipolar sources [10; 26; 56; 57; 58; 59; 60; 61; 62; 63; 64; 65; 66; 67, 68; 69]. The source localizations results have then been evaluated by comparing them either to those of other clinical, structural or functional imagery exams, to the outcome after surgical resection of the suspected area, or in some cases to the results to intracranial recordings from subdural and/or depth electrodes. In a few studies simultaneous recordings of intra- and extracranial epileptiform activity have been used, permitting direct comparison of epileptic events that occur simultaneously on the implanted and surface electrodes. The latter technique has been used in combination with dipole modeling to localize mesiotemporal sources or to differentiate between mesial and lateral temporal sources in patients with temporal lobe epilepsy [70; 71; 72; 73]. However, since the interpretations about source localizations in dipole studies are based rather on the orientation than on the location of the dipoles [3], the estimation of the focus lcoation with these techniques is only indirect.

Other alternative approaches that have been suggested for epileptic focus localization include EPIFOCUS, which determines the probability of finding a focal source at any given point in a discrete solution space [44; 55; 74; 75]. Also the distributed source models discussed above have been repeatedly applied to epileptic data with considerable success [14; 17; 29; 42; 74; 76; 77; 78].

Two recent studies by our group have assessed the clinical yield of ESI in localizing interictal epileptiform activity, and compared it to that of other neuroimaging techniques. In the first study [4] a group of 32 patients with intractable epilepsy of different origins in whom a presurgical workup had identified a focal epileptogenic region, were investigated with 125 EEG recording electrodes. The ESI using the EPIFOCUS method correctly localized the focus in 94% of the patients. In the second study [51] standard 19-29 electrode EEG recordings were analyzed in 30 children with focal epilepsy of different locations. In this study we applied the statistical approach described above based on a distributed linear inverse solution (weigthed minimum norm). We found 100% correct localization for the extratemporal patients (100% correct), whereas the temporal lobe patients showed larger localization errors (77% correct localizations). It could be demonstrated that the reason for the less precise temporal lobe results in this study was the low number of recording electrodes, and the importance of sufficient spatial sampling was emphasized. In both studies the clinical yield with ESI (in terms of correct localization of the epileptic focus) was similar or superior to SPECT and PET

### Propagation of Interictal Epileptiform Activity

It has been demonstrated in several studies that propagation of the epileptic activity may occur, both within the temporal lobe [28; 79; 80; 81; 82] and between distant areas [80; 83;

84]. Several possible intracerebral connections exist, which might play a role in the propagation both of interictal epileptiform activity and of epileptic seizures. Examples of such connections are from hippocampus/amygdala to the neocortices [85; 86], and from the amygdala to the temporal pole, the superior temporal gyrus, and the medial and orbital surfaces of the frontal lobe [87]: Projections also exist between amygdala and thalamus, between thalamus and all areas of prefrontal cortex, and between anterior cingulate and orbitofrontal cortex and the amygdala. (see [88] and references). Propagation between the temporal lobes might occur through the hippocampal commisure and between the frontal lobes through callosal projections [89]. From the occipital lobe propagation by several different pathways medially and laterally, above and below the sylvian fissure, has been predicted [90].

Source localization studies of surface EEG have suggested different propagation pathways during the course of a spike, and these pathways have in some cases been verified through investigations with intracranial electrodes. Thus, in some studies [3; 60; 62], sequential activation of two different dipole sources have been explained in terms of activation of mesiobasal and lateral temporal neocortex respectively, an interpretation which has found support in intracranial investigations [73; 81; 91]. Between the temporal tip and anterolateral temporal neocortex, intracranially verified propagation has been demonstrated both with spatiotemporal dipole modelling [20], and with a current density algorithm [14]. Another cortical current density study [92] showed propagation before the spike peak from a lateral temporal lesion area to both anterior, posterior and partly contralateral temporal regions. In [10] three cases were presented (with a temporal, frontal, and parietal focus respectively), in all of which propagated sources dominated at the timepoint of the most prominent scalp peak. In another study [75] propagation outside the subsequently resected area was also demonstrated for several patients (figure 3). Other examples of EEG and MEG studies where the results have been interpreted in terms of epileptic source propagation are [80; 82; 84; 92].

Spike propagation may also occur through sequential activation of an increasing number of brain areas. From intracranial recordings [79; 93] a significant degree of neuronal recruitment has been demonstrated as discharges propagate, resulting in larger, delayed and less synchronized spikes outside the pacemaker zone [79]. In a study where spatiotemporal dipole modelling of surface recorded epileptic spikes was used [28] the results suggested consecutive activation of several different generators, leading to progressively more complex activation patterns as the spike evolved.

In conclusion, these different studies indicate: (i) that fast propagation of interictal epileptiform activity does occur, (ii) that areas both in the vicinity and at a distance from the focus may be involved in the propagation process, and (iii) that it is indeed possible to retrieve these intracerebral propagation patterns on the surface and localize the sources with ESI.

It is of particular importance to consider interictal spike propagation when the timepoint to use for source localizations shall be decided. In many previous EEG and MEG studies [41; 64; 94; 95] the spike peak has been used for the source localizations. However, in view of the fact that propagated sources may very well dominate at the spike peak, it has been suggested that the ascending phase of a spike or a sharp wave would be a more appropriate timepoint for source localization [10; 75; 92].

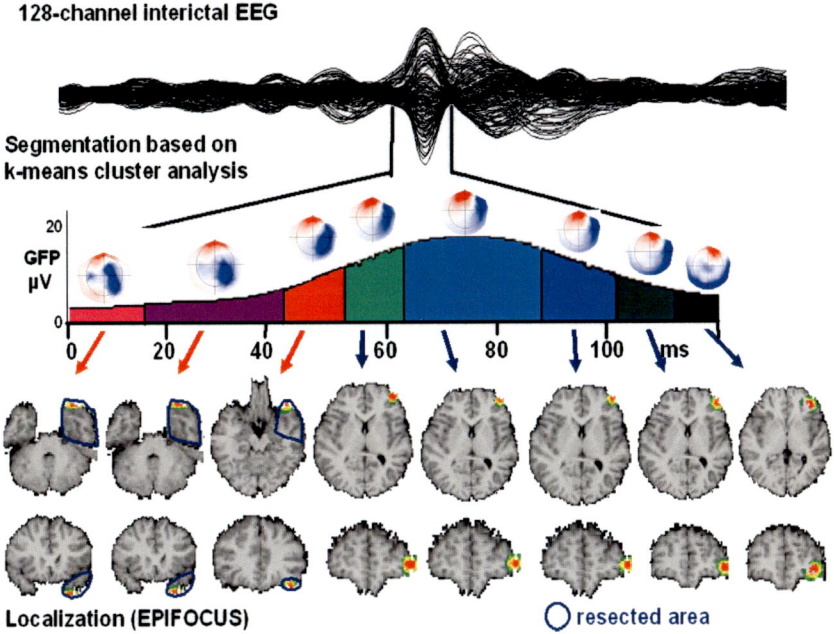

Figure 3. Propagation of the epilpetiform activity during the course of an epileptic spike recorded with 128 electrodes (top). Through a temporal segmentation procedure based on a k-means cluster analysis periods during the spike showing a relatively stable voltage map configuration are identified. A source localization algorithm (EPIFOCUS) is used to localize the source of activity during each stable period. The patients suffers from a right temporal lobe epilepsy and is seizure free after temporal lobe resection. During the early phases of the spike the ESI localizes the source to the right temporal lobe. During the later phases of the spike, however, the ESI shows propagation of the epileptiform activity to the right frontal lobe.

## D. CHARACTERIZATION OF ICTAL ONSET AND SPREAD

### Electrophysiological Techniques to Quantitatively Determine Ictal Onset and Spread

Different methods to quantitatively determine the origin and spread of ictal epileptiform activity have been suggested, both for extra- and intracranially recorded seizures. Examples of such techniques are autoregressive modelling [96; 97], wavelet analysis (for a review see [98]), and methods related to non-linear dynamics [99; 100; 101]. In temporal lobe patients spatiotemporal dipole modelling (102) has been used to localize surface recorded ictal rhythms [70; 103]. Several studies have also shown that important information about the type and the spread of ictal epileptiform activity can be obtained from the spectral properties of the ictal EEG [104; 105; 106; 107]. A special variant of this approach is the so called FFT approximation technique [108] which can be used to study the spatial distribution of the activity in a certain frequency instead of the spectral properties of individual channels. With time-frequency analysis [109; 110], the temporal resolution of the frequency analysis is improved down to a millisecond time scale. Temporal segmentation of EEG patterns [111; 112; 113; 114; 115] has been used in epilepsy in order to classify initiation and spread of the seizures [97; 116; 117]. A further development of the latter technique is the concept of "functional microstates" [118], which allows for temporal segmentation of whole map series instead of single channel data.

## Temporal Lobe Seizures

From intracranial studies it has been demonstrated that seizures originating in the amygdala and parahippocampal gyrus are more likely to propagate than seizures originating in the hippocampus, to some extentd due to stronger inhibitory circuits in the latter structure [119; 120]. In analogy, secondary propagation from mesiotemporal structures to other areas is more rapid if the primary focus is neocortical [119]. Within the temporal lobe, Bartolomei et al [121; 122] have described four different propagation patterns between mesial and neocortical structures, mesial onset with fast and slow neocortical propagation, and neocortical onset with fast and slow mesial propagation. In [89] and [119] different propagation pathways between the temporal and frontal lobes ipsi- and contralateral have been described.

The extracranial ictal patterns of temporal lobe seizures are quite well characterized. One frequently encountered pattern is an initial bilateral flattening followed by a gradual build-up of lateralized rhythmic activity [123]. Such a pattern of lateralized 5 Hz or higher frequency activity (called initial focal) is considered to be highly predictive of mesial temporal ictal onset [73; 124; 125; 126; 127; 128]. Another pattern of lateralized rhythmic activity preceded by a diffuse electrographical change (delayed focal) has also been found to be quite predictive of mesial temporal seizure onset [127]. Seizures with slower rhythmic activity, as well as non-lateralized or bilateral ictal patterns, have been attributed to seizures of neocortical temporal lobe origin [73; 124; 126; 129; 130].

It has been possible to use dipole source localization procedures to quantitatively differentiate mesiotemporal from neocortical ictal patterns. In [62] the early ictal rhythms of mesiotemporal seizures could be modeled using a strong vertical dipole component, whereas lateral temporal neocortical seizures were best modeled with radial dipoles. Boon et al. [59] found similar differences when comparing seizures in patients with temporal (mostly medial) lesions compared to patients with extratemporal neocortical lesions. Through a full-scalp frequency analysis it is possible to define the dominant global frequency in the ictal EEG, and to determine the beginning of power increase in this dominant frequency around seizure onset [17; 108]. In two studies [131; 132] dipole localization of the dominant frequency have been shown to give correct and reproducible source localization results for seizures originating in the mesial temporal area. In a recent study, which is in preparation to be published, we suggest an FFT approach, which provides a spatio-temporal characterization of epileptic seizures, while at the same time identifying the different areas in the brain that are active during seizure development. The results in 3 patients with different focus location within the temporal lobe are shown in figure 4.

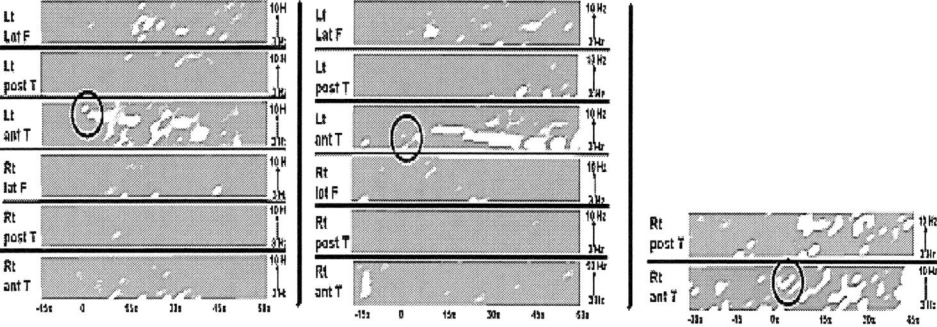

Figure 4. Source localizations (EPIFOCUS) of one seizure in each of three patients, Pat1 with a mesiotemporal ictal onset (left), Pat2 with a neocortical ictal onset (centre), and Pat3 with intracranially verified propagation from anterior to posterior temporal regions (right). The FFT-approximation is used to decompose the ictal activity is into its frequency components, and for each timeframe of 2 seconds, a distributed source model is used to determine the power of each frequency between 3 and 10 Hz within different regions of interest (ROI:s) in the brain. In Pat1 with the mesiotemporal onset, a 9-10 Hz activity confined to the left anterior temporal ROI is seen around seizure onset. During the first 10 seconds after ictal onset the activity shifts to a broader activation of several frequencies in the 4-7 Hz range, but still mailnly confined to the same region. In Pat2 with the neocortical ictal onset the same region as in Pat1 is activated, but with a rather different frequency profile (onset 3-4 Hz with subsequent shift to higher frequencies). The different frequencies at ictal onset for mesiotemporal and neocortical temporal foci is in line with what has been described in the literature (Ebersole and Pacia, 1996). In Pat3 ictal onset is on the right anterior temporal region but with fast subsequent activation of the right posterior temporal ROI. The propagation from anterior to posterior temporal areas has been confirmed with intracranial electrodes. Lt=left, Rt=right, ant=anterior, lat=lateral, post=posterior, F=frontal, T=temporal.

## Extratemporal Seizures

Neocortical seizures are far more variable concerning frequency content and morphology [133]. Some studies have found the so called electrodecremental pattern (low voltage high frequency discharges with a general low voltage background) to be the most characteristic onset pattern in these cases [73; 133], and there is a clear similarity between this pattern and the mesiotemporal onset pattern (b) described above. Another possible pattern of neocortical temporal ictal onset is a repetitive <5Hz sharp wave activity, which differs from the mesiotemporal onset patterns (a) described above by being much less focal. For parietal seizures ictal propagation has been observed from the parieto-temporal region to the frontal and parietal operculum, from the posterior parietal to the medial frontal region, and from the superior lobule of the parietal lobe to the temporal lobe on the same side [134]. Occipital lobe seizures have been shown to spread from the supero-lateral occipital lobe towards the temporal convexity, from inferomedial occipital cortex to the rhinencephalon, and from the supracalcarine cortex to the parietal lobe [135; 136]. In children rapid propagation to the opposite occipital lobe through the corpus callosum may occur [136].

Frontal lobe seizures are often difficult to asses on surface EEG, due to the inaccessibility of much of the lobes to surface electrodes, widespread anatomical connections to other areas, frequent bilateral foci, and variable size of the ictal onset zone [137]. Even so correctly localized patterns (visually estimated), often in the form of rhythmic epileptiform activity, has been reported in as many as 65% of cases with lateral frontal lobe epilepsy [126]. In cases with mesial frontal sources the proportion of focal patterns was considerably lower (24%). The ictal patterns of parietal and occipital sources are quite variable, and the seizures show a strong tendency to propagate, typically to the temporal or frontal lobes or to the opposite side [138; 139; 140]. Even so, adequately localizing surface ictal patterns (again visually estimated), mainly in the form of paroxysmal fast or repetitive activity, have been obtained in around 50% of patients with lateral parietal sources [126]. For occipital sources lateralized patterns, mainly rhythmic delta, have been obtained in approximately the same proportion as for parietal sources [126].

Studies using ESI of ictal data in extratemporal epilepsy are to our knowledge quite few. [78] found localizations that corresponded well to MR lesion in patients with a frontal lobe focus. In a recent study [52] used a distributed source model (LORETA [13]) to localize the source of ictal epileptiform activity in three patients with partial status epilepticus, and found the results to correlate well with the results of ictal PET in the same patients. In [141] the same distributed source model was used to analyze spike-wave activity in absence seizures, and it was suggested that absence seizures are not truly "generalized," with immediate global cortical involvement, but rather involve selective cortical networks, including orbital frontal and mesial frontal regions, in the propagation of ictal discharges [142; 143; 144]. Another illustration of the possibility to use ESI to visualize fast propagation in frontal lobe seizures is given in figure 5, which is from the same study as figure 4.

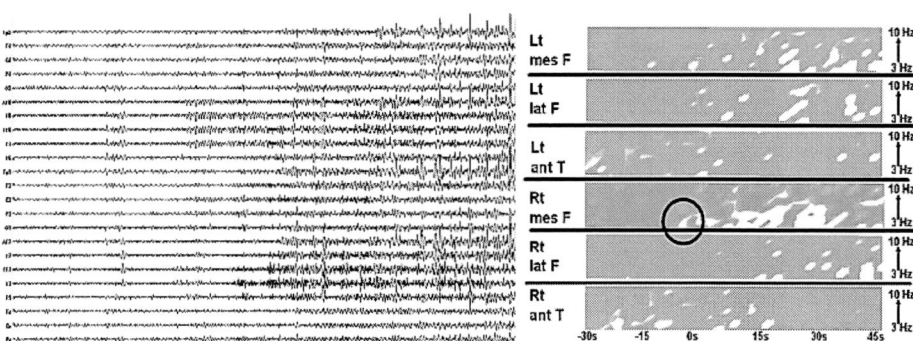

Figure 5. Source localization (EPIFOCUS) in one patient with a frontal lobe focus. For details about the analysis see figure 4. The seizure starts with a 3-4 Hz activity in the right mesial frontal region and propagation is seen within a few seconds mainly to contralateral frontal areas.

## CONCLUSION

We conclude that Electrical Source Imaging (ESI) is a useful additional tool in the presurgical workup of patients with pharmaco resistant partial epilepsy. For optimal source localization accuracy it is important to use several different source models, to perform the source reconstructions in a realistic head model, preferably the MRI of the individual patient, and that a sufficient number of recording electrodes (at least 60-70) are used. The results can also be improved by the use of statistical SPM like techniques to determine areas of significant activation. With ESI the focus of interictal and ictal epileptiform activity can be characterized, both in terms of focus localization and in terms of visualizing complex propagation patterns. To accurately determine the anatomical origin of interictal and ictal epileptiform activity from EEG recordings is of crucial importance in presurgical epilepsy investigations, and the possibility to characterize interictal and ictal propagation patterns also gives important additional information for understanding the epileptic process in the individual patient, for instance when it comes to understanding discrepancies between ESI findings and ictal semiology or SPECT/PET/fMRI findings.

## ACKNOWLEDGMENT

Supported by the Centre d'Imagerie BioMedicale (CIBM) of Geneva and Lausanne.

## REFERENCES

[2] Rosenow F, Lüders H. 2001. Presurgical evaluation of epilepsy. *Brain.* 124:1683-1700.
[3] Binnie CD, Stefan H. 1999. Modern electroencephalography: its role in epilepsy management. *Clinical Neurophysiology.* 110:1671-1697.
[4] Ebersole JS. 1997. Defining epileptogenic foci: Past Present Future. *Journal of Clinical Neurophysiology.* 14:470-483.
[5] Michel CM, Lantz G, Spinelli L, Grave de Peralta Menedez R, Landis T, Seeck M. 2004a. 128-channel EEG source imaging in epilepsy: clinical yield and localization precision. *Journal of Clinical Neurophysiology.* 21(2):71-83.
[6] Anogianakis G, Badier JM, Barrett G, Erné S, Fenici R, Fenwick P, Grandori F, Hari R, Ilmoniemi R, Mauguière F, Lehmann D, Perrin F, Peters M, Romani G-L, Rossini PM. 1992. A consensus statement of relative merits of EEG and MEG. *Electroencephalography and Clinical Neurophysiology.* (Editorial) 82:317-319.
[7] Barkley GL, Baumgartner C. 2003. MEG and EEG in epilepsy. *Journal of Clinical Neurophysiology.* 20(3):163-178.
[8] Liu AK, Dale AM, Belliveau JW. 2002. Monte Carlo simulation studies of EEG and MEG localization accuracy. *Human Brain Mapping.* 16:47-62.
[9] Malmivuo J, Suihko V, Eskola H. 1997. Sensitivity distributions of EEG and MEG measurements. *IEEE Transactions on Biomedical Engineering.* 44:196-208.
[10] Wikswo JPJ, Gevins A, Williamson SJ. 1993. The future of EEG and MEG. *Electroencephalography and Clinical Neurophysiology.* 87:1-9.
[11] Scherg M, Bast T, Berg P. 1999. Multiple source analysis of interictal spikes: goals, requirements, and clinical value. *Journal of Clinical Neurophysiology.* 16(3):214-224.
[12] Hämäläinen MS, Ilmoniemi RJ. 1984. Interpreting measured magnetic fields of the brain: estimation of current distributions. Helsinki: Helsinki University of Technology. Report nr TKK-F-A559. 28p p.
[13] Lawson CL, Hanson RJ. 1974. Solving least square problems. Englewool Cliffs New Jersey: Prentice-Hall.

[14] Pascual-Marqui RD, Michel CM, Lehmann D. 1994. Low resolution electromagnetic tomography: a new method for localizing electrical activity in the brain. *International Journal of Psychophysiology.* 18:49-65.

[15] Fuchs M, Wagner M, Köhler T, Wischmann H-A. 1999. Linear and nonlinear current density reconstructions. *Journal of Clinical Neurophysiology.* 16:267-295.

[16] George JS, Aine CJ, Mosher JC, Schmidt DM, Ranken DM, Schlitt HA, Wood CC, Lewine JD, Sanders JA, Belliveau JW. 1995. Mapping function in the human brain with magnetoencephalography, anatomical magnetic resonance imaging, and functional magnetic resonance imaging. *Journal of Clinical Neurophysiology.* 12:406-431.

[17] Hämäläinen M, Hari R, Ilmoniemi RJ, Knuutila J, Lounesmaa OV. 1993. Magneto-encephalography - theory, instrumentation, and applications to noninvasive studies of the working human brain. *Rev. Mod. Phys.* 65:413-497.

[18] Michel CM, Grave de Peralta R, Lantz G, Gonzalez Andino S, Spinelli L, Blanke O, Landis T, Seeck M. 1999. Spatio-temporal EEG analysis and distributed source estimation in presurgical epilepsy evaluation. *Journal of Clinical Neurophysiology.* 16:225-238.

[19] Mosher JC, Spencer ME, Leahy RM, Lewis PS. 1993. Error bounds for EEG and MEG dipole source localization. *Electroencephalography and Clinical Neurophysiology.* 86:303-321.

[20] Ebersole JS. 1999. EEG source modeling. The last word. *Journal of Clinical Neurophysiology.* 16(3):297-302.

[21] Ary JP, Klein SA, Fender DH. 1981. Location of sources of evoked scalp potentials: corrections for skull and scalp thicknesses. *IEEE Transactions on Biomedical Engineering.* 128:447-452.

[22] Berg P, Scherg M. 1994. A fast method for forward computation of multiple-shell spherical head models. *Electroencephalography and Clinical Neurophysiology.* 90:58-64.

[23] De Munck JC, Peters MJ. 1993. A fast method to compute the potential in the multisphere model. *IEEE Transactions on Biomedical Engineering.* 40:1166-1174.

[24] Zhang Z, Jewett DL. 1993. Insidious errors in dipole localization parameters at a single time-point due to model misspecification of number of shells. *Electroencephalography and Clinical Neurophysiology.* 88:1-11.

[25] De Munck JC, Van Dijk B, Wand Spekreijse H. 1988. Mathematical dipoles are adequate to describe realistic generators of human brain activity. *IEEE Transactions on Biomedical Engineering.* 35:960-966.

[26] Zhang Z. 1995. A fast method to compute surface potentials generated by dipoles within multilayer anisotropic spheres. *Physics in Medicine and Biology.* 40:335-349.

[27] Dieckmann V, Becker W, Jürgens R, Kleiser B, Richter HP, Wollinsky KH. 1998. Localisation of epileptic foci with electric, magnetic and combined electromagnetic models. *Electroencephalography and Clinical Neurophysiology.* 106:297-313.

[28] Lagerlund TD, Sharbrough FW, Jack CRJ, Erickson BJ, Strelow DC, Cicora KM, Busacker NE. 1993. Determination of 10-20 system electrode locations using magnetic resonance image scanning with markers. *Electroencephalography and Clinical Neurophysiology.* 86:7-14.

[29] Merlet I, Garcia-Larrea L, Gregoire MC, Lavenne F, Mauguière F. 1996. Source propagation of interictal spikes in temporal lobe epilepsy. Correlations between spike dipole modelling and [18F]fluorodeoxyglucose PET data. *Brain.* 119(Apr):377-392.

[30] Seri S, Cerquiglini A, Pisani F, Michel CM, Pascual Marqui RD, Curatolo P. 1998. Frontal lobe epilepsy associated with tuberous sclerosis: electroencephalographic-magnetic resonance image fusioning. *Journal of Child Neurology.* 13:33-38.

[31] Towle VL, Bolanos J, Suarez D, Tan K, Grzeszczuk R, Levin DN, Cakmur R, Frank SA, Spire JP. 1993. The spatial location of EEG electrodes: locating the best-fitting sphere relative to cortical anatomy. *Electroencephalography and Clinical Neurophysiology.* 86:1-6.

[32] Yoo SS, Guttmann CR, Ives JR, Panych LP, Kikinis R, Schomer DL, Jolesz FA. 1997. 3D localization of surface 10-20 EEG electrodes on high resolution anatomical MR images. *Electroencephalography and Clinical Neurophysiology.* 102:335-339.

[33] Hämäläinen M, Sarvas J. 1989. Realistic conductor geometry model of the human head for interpretation of neuromagnetic data. *IEEE Transactions on Biomedical Engineering.* 36:165-171.

*[34]* Bertrand O, Thevenet M, Perrin F. 1991. Finite element method in brain electrical activity studies. In: Nenoner J, Rajala HM, Katila T, editors. *Biomagnetic localization and 3D modeling. Otaniemi.*

[35] Miller CE, Henriquez CS. 1990. Finite element analysis of bioelectric phenomena. *Critical Reviews in Biomedical Engineering.* 18:207-233.

[36] Yan Y, Nunez PL, Hart RT. 1991. Finite-element model of human head: scalp potentials due to dipole sources. *Medical and Biological Engineering and Computing.* 29:475-481.

[37] Spinelli L, Gonzalez Andino S, Lantz G, Michel CM. 2000. Electromagnetic Inverse Solutions in Anatomically Constrained Spherical Head Models. *Brain Topography.* 13:115-125.

[38] Babiloni F, Carducci F, Cincotti F, Del Gratta C, Roberti GM, Romani GL, Rossini PM, Babiloni C. 2000. Integration of high resolution EEG and functional magnetic resonance in the study of human movement-related potentials. *Methods of Information in Medicine.* 39:179-182.

[39] Gevins A. 1990. Distributed neuroelectric patterns of human neocortex during simple cognitive tasks. *Progress in Brain Research.* 85:337-345.

[40] Srinivasan R, Nunez PL, Tucker DM, Silberstein RB, Cadusch PJ. 1996. Spatial sampling and filtering of EEG with spline laplacians to estimate cortical potentials. *Brain Topography.* 8:355-366.

[41] Herrendorf G, Steinhoff BJ, Kolle R, Baudewig J, Waberski TD, Buchner H, Paulus W. 2000. Dipole-source analysis in a realistic head model in patients with focal epilepsy. *Epilepsia.* 41:71-80.

[42] Huppertz HJ, Hof E, Klisch J, Wagner M, Lucking CH, Kristeva-Feige R. 2001b. Localization of interictal delta and epileptiform EEG activity associated with focal epileptogenic brain lesions. *Neuroimage.* 13(1):15-28.

[43] Waberski TD, Gobbelé R, Herrendorf G, Steinhoff BJ, Kolle R, Fuchs M, Paulus W, Buchner H. 2000. Source reconstruction of mesial temporal epileptiform activity: comparison of inverse techniques. *Epilepsia.* 41:1574-1583.

[44] Lantz G, Grave de Peralta R, Spinelli L, Seeck M, Michel CM. 2003a. Epileptic source localization with high density EEG: how many electrodes are needed? *Clinical Neurophysiology.* 114(1):63-69.

[45] Grave de Peralta R, Gonzalez S, Lantz G, Michel CM, Landis T. 2001. Noninvasive localization of electromagnetic epileptic activity. I. Method descriptions and simulations. *Brain Topography.* 14:131-137.

[46] Tucker DM. 1993. Spatial sampling of head electrical fields: the geodesic sensor net. *Electroencephalography and Clinical Neurophysiology.* 87:154-163.

[47] Park HJ, Kwon JS, Youn T, Pae JS, Kim JJ, Kim MS, Ha KS. 2002. Statistical parametric mapping of LORETA using high density EEG and individual MRI: application to mismatch negativities in schizophrenia. *Human Brain Mapping.* 17(3):168-178.

[48] Fallgatter AJ, Bartsch AJ, Zielasek J, Herrmann MJ. 2003. Brain electrical dysfunction of the anterior cingulate in schizophrenic patients. *Psychiatry Research.* 30:37-48.

[49] Kounios J, Smith RW, Yang W, Bachman P, D'Esposito M. 2001. Cognitive association formation in human memory revealed by spatiotemporal brain imaging. *Neuron.* 29(1):297-306.

[50] Michel CM, Seeck M, Murray MM. 2004c. The speed of visual cognition. *Supplements to Clinical Neurophysiology* 57:617-627.

[51] Murray MM, Michel CM, Grave de Peralta R, Ortigue S, Brunet D, Gonzalez Andino S, Schnider A. 2004. The sound and the memory: rapid, incidental discrimination of visual and multisensory memories. *Neuroimage.* 21:125-135.

[52] Sperli F, Spinelli L, Seeck M, Kurian M, Michel CM, Lantz G. 2006. EEG source imaging in pediatric epilepsy surgery: a new perspective in presurgical workup. *Epilepsia.* 47(6):981-990.

[53] Zumsteg D, Wennberg RA, Treyer V, Buck A, Wieser HG. 2005b. H2(15)O or 13NH3 PET and electromagnetic tomography (LORETA) during partial status epilepticus. *Neurology.* 65(10):1657-1660.

[54] Baumgartner C. 2004. Controversies in clinical neurophysiology. MEG is superior to EEG in the localization of interictal epileptiform activity: Con. *Clinical Neurophysiology.* 115(5):1010-1020.

[55] Ebersole JS. 2000. Noninvasive localization of epileptogenic foci by EEG source modeling. *Epilepsia.* 41(suppl 3):24-33.

[56] Michel CM, Murray M, Lantz G, Gonzalez S, Spinelli L, Grave de Peralta R. 2004b. *EEG source imaging. Clinical Neurophysiology.* 115(10):2195-2222.

[57] Barth DS, Sutherling W, Engel J, Beatty J. 1982. Neuromagnetic localization of epileptiform spike activity in the human brain. *Science.* 218:891-894.

[58] Baumgartner C, Lindinger G, Ebner A, Aull S, Serles W, Olbrich A, Lurger S, Czech T, Burgess R, Luders H. 1995. Propagation of interictal epileptic activity in temporal lobe epilepsy. *Neurology.* 45:118-122.

[59] Boon P, D'Have M. 1995. Interictal and ictal dipole modelling in patients with refractory partial epilepsy. *Acta Neurologica Scandinavica.* 92:7-18.

[60] Boon P, D'Have M, Adam C, Vonck K, Baulac M, Vandekerckhove T, De Reuck J. 1997. Dipole modeling in epilepsy surgery candidates. *Epilepsia.* 38(2):208-218.

[61] Ebersole JS. 1991. EEG dipole modelling in complex partial epilepsy. *Brain Topography.* 4:113-123.

[62] Ebersole JS. 1992. Equivalent dipole modelling - a new EEG method for localisation of epileptogenic foci. In: Pedley TA, Meldrum BS, editors. *Recent Advances in Epilepsy.* London: Churchill Livingston. p. 51-71.

[63] Ebersole JS. 1994. Non-invasive localization of the epileptogenic focus by EEG dipole modeling. *Acta Neurologica Scandinavica Suppl.* 152:20-28.

[64] Ebersole JS, Wade PB. 1991. Spike voltage topography identifies two types of frontotemporal epileptic foci. *Neurology.* 41:1425-1433.

[65] Lantz G, Ryding E, Rosén I. 1994. Three dimensional localisation of interictal epileptiform activity with dipole analysis: comparison with intracranial recordings and SPECT findings. *J. Epilepsy.* 7:117-129.

[66] Nakasato N, Levesque MP, Barth DS, Baumgartner C, Rogers RL, Sutherling WW. 1994. Comparisons of MEG, EEG, ECoG source localization in neocortical epilepsy in humans. *Electroencephalography and Clinical Neurophysiology.* 91:171-172.

[67] Shindo K, Ikeda A, Musha T, Terada K, Fukuyama H, Taki W, Kimura J, Shibasaki H. 1998. Clinical usefulness of the dipole tracing method for localizing interictal spikes in partial epilepsy. *Epilepsia.* 39:371-379.

[68] Stefan H, Schneider S, Feistel H, Pawlik G, Schüler P, Abraham-Fuchs K, Schegel T, Neubauer U, Huk WJ. 1992. Ictal and interictal activity in partial epilepsy recorded with multichannel magnetoencephalography: correlation of electroencephalography/ electrocorticography, magnetic resonance imaging, single photon emission computed tomography, and positron emission tomography findings. *Epilepsia.* 33:874-887.

[69] Stefan H, Schuler P, Abraham-Fuchs K, Schneider S, Gebhardt M, Neubauer U, Hummel C, Huk WJ, Thierauf P. 1994. Magnetic source localization and morphological changes in temporal lobe epilepsies: comparison of MEG/EEG, ECoG and volumetric MRI in presurgical evaluation of operated patients. *Acta Neurologica Scandinavica Supplementum.* 152:83-88.

[70] Wong PKH. 1991. Source modelling of the rolandic focus. *Brain Topography.* 4:105-112.

[71] Boon P, D'Havé M, Adam C, Vonck K, Baulac M, Vandekerckhove T, De Reuck J. 1996. Dipole modeling in epilepsy surgery candidates. *Epilepsia.* 38:208-218.

[72] Lantz G, Ryding E, Holub M, Rosén I. 1996. Simultaneous intracranial and extracranial recording of interictal epilptiform activity in patients with drug resistant partial epilepsy - patterns of conduction and results from dipole reconstruction. *Electroencephalography and Clinical Neurophysiology.* 99:69-78.

[73] Merlet I, Garcia-Larrea L, Ryvlin P, Isnard J, Sindou M, Mauguière F. 1998. Topographical reliability of mesio-temporal sources of interictal spikes in temporal lobe epilepsy. *Electroencephalography and Clinical Neurophysiology.* 107:206-212.

[74] Pacia SV, Ebersole JS. 1997. Intracranial EEG substrates of scalp ictal patterns from temporal lobe foci. *Epilepsia.* 38:642-654.

[75] Lantz G, Grave de Peralta R, Gonzalez S, Michel CM. 2001a. Noninvasive localization of electromagnetic epileptic activity. II. Demonstration of sublobar accuracy in patients with simultaneous surface and depth recordings. *Brain Topography.* 14:139-147.

[76] Lantz G, Spinelli L, Seeck M, Grave de Peralta Menendez R, Sottas C, Michel C. 2003b. Propagation of interictal epileptiform activity can lead to erroneous source localizations: A 128 channel EEG mapping study. *Journal of Clinical Neurophysiology.* 20(5):311-319.

[77] Lantz G, Michel CM, Pasqual-Marqui R, Spinelli L, Seeck M, Seri S, Landis T, Rosén I. 1997a. Extracranial localisation of intracranial interictal epileptiform activity using LORETA (Low Resolution Electromagnetic Tomography). *Electroencephalography and Clinical Neurophysiology.* 102:414-422.

[78] Lantz G, Spinelli L, Grave de Peralta R, Seeck M, Michel CM. 2001b. Localisation de sources distribuées et comparaison avec l'IRM fonctionnelle. *Epileptic Disorders.* 3(Special issue):45-58.

[79] Worrell GA, Lagerlund TD, Sharbrough FW, Brinkmann BH, Busacker NE, Cicora KM, O'Brien TJ. 2000. Localization of the epileptic focus by low-resolution electromagnetic tomography in patients with a lesion demonstrated by MRI. *Brain Topography.* 12(4):273-282.

[80] Alarcon G, Seoane JJG, Binnie CD, Miguel MCM, Juler J, Polkey CE, Elwes RD, Blasco JMO. 1997. Origin and propagation of interictal discharges in the acute electrocorticogram. Implications for pathophysiology and surgical treatment of temporal lobe epilepsy. *Brain.* 120(Dec):259-282.

[81] Emerson RG, Turner CA, Pedley TA, Walczak TS, Forgione M. 1995. Propagation patterns of temporal spikes. *Electroencephalography and Clinical Neurophysiology.* 94(5):338-348.

[82] Merlet I, Gotman J. 1999. Reliability of dipole models of epileptic spikes. *Clinical Neurophysiology.* 110:1013-1028.

[83] Sutherling WW, Barth DS. 1989. Neocortical propagation in temporal lobe spike foci on magnetoencephalography and electroencephalography. *Annals of Neurology.* 25:373-381.

[84] Alarcon G, Guy CN, Binnie CD, Walker SR, Elwes RDC, Polkey CE. 1994. Intracerebral propagation of interictal activity in partial epilepsy: implications for source localisation. *Journal of Neurology, Neurosurgery and Psychiatry.* 57:435-449.

[85] Leal AJ, Passao V, Calado E, Vieira JP, Silva Cunha JP. 2002. Interictal spike EEG source analysis in hypothalamic hamartoma epilepsy. *Clinical Neurophysiology.* 113(12):1961-1969.

[86] Lopes da Silva FH, Witter MP, Boejinga PH, Lohman AHM. 1990. Anantomic organisation and physiology of the limbic cortex. *Physiological Reviews.* 70:453-511.

[87] Pandya DN, Yeterian EH. 1987. Hodology of limbic and related structures: cortical and commissural connections. In: Wieser HG, Elger CE, editors. *Presurgical evaluation of epileptics.* Berlin: Springer-Verlag. p. 3-14.

[88] Amaral DG, Price JL. 1984. Amygdalo-cortical projections in the monkey (Macaca fascilularis). *Journal of Comparative Neurology.* 230:465-496.

[89] Lieb JP, Dasheiff RM, Engel JJ. 1991. Role of the frontal lobes in the propagation of mesial temporal lobe seizures. *Epilepsia.* 32(6):822-837.

[90] Adam C, Sain-Hilaire JM, Richer F. 1994. Temporal and spatial characteristics of intracerebral seizure propagation: predicitive value in surgery for temporal lobe epilepsy. *Epilepsia.* 35(5):1065-1072.

[91] Ajmone-Marsan C, Ralston BL. 1957. *The epileptic seizure. Its functional morphology and diagnostic significance.* Thomas CC, editor: Springfield.

[92] Ebersole JS, Hawes S, Scherg M. 1995. Intracranial EEG validation of spike propagation predicted by dipole models. *Electroencephalography and Clinical Neurophysiology.* 95:18.

[93] Huppertz HJ, Hoegg S, Sick C, Lucking CH, Zentner J, Schulze Bonhage A, Kristeva-Feige R. 2001a. Cortical current density reconstruction of interictal epileptiform activity in temporal lobe epilepsy. *Clinical Neurophysiology.* 112(9):1761-1772.

[94] Hufnagel A, Dumpelmann M, Zentner J, Schijns O, Elger CE. 2000. Clinical relevance of quantified intracranial interictal spike activity in presurgical evaluation of epilepsy. *Epilepsia.* 41(4):467-478.

[95] Iwasaki M, Nakasato N, Shamoto H, Nagamatsu K, Kanno A, Hatanaka K, Yoshimoto T. 2002. Surgical implications of neuromagnetic spike localization in temporal lobe epilepsy. *Epilepsia.* 43(4):415-424.

[96] Lantz G, Ryding E, Rosén I. 1997b. Dipole reconstruction as a method for identifying patients with mesolimbic epilepsy. *Seizure.* 6:303 - 310.

[97] Franaszczuk PJ, Bergey GK, Kaminski MJ. 1994. Analysis of mesial temporal seizure onset and propagation using the directed transfer function method. *Electroencephalography and Clinical Neurophysiology.* 91:413-427.

[98] Gath I, Feuerstein C, Pham DT, Rondouin G. 1992. On the tracking of rapid dynamic changes in seizure EEG. *IEEE Transactions on Biomedical Engineering.* 39:952-958.

[99] Unser M, Aldroubi A. 1996. A review of wavelets in biomedical applications. *Proc. IEEE.* 84:626-638.

[100] Bullmore ET, Brammer MJ, Bourlon P, Alarcon G, Polkey CE, Elwes R, Binnie CD. 1994. Fractal analysis of electroencephalographic signals intracerebrally recorded during 35 epileptic seizures: evaluation of a new method for synoptic visualisation of ictal events. *Electroencephalography and Clinical Neurophysiology.* 91:337-345.

[101] Lehnertz K, Elger CE. 1995. Spatio-temporal dynamics of the primary epileptogenic area in temporal lobe epilepsy characterized by neuronal complexity loss. *Electroencephalography and Clinical Neurophysiology.* 95:108-117.

[102] Pijn JP, Van Neerven J, Noest A, Lopes da Silva FH. 1991. Chaos or noise in EEG signals; dependence on state and brain site. *Electroencephalography and Clinical Neurophysiology.* 79:371-381.

[103] Scherg M. 1990. Fundamentals of dipole source potential analysis. In: Grandori F, Hoke M, Romani GL, editors. *Auditory evoked magnetic fields and electric potentials.* Basel: Karger. p. 40-96.

[104] Assaf BA, Ebersole JS. 1997. Continuous source imaging of scalp ictal rhythms in temporal lobe epilepsy. *Epilepsia.* 38:1114-1123.

[105] Alarcon G, Binnie CD, Elwes RDC, Polkey CE. 1995. Power spectrum and intracranial EEG patterns at seizure onset in partial epilepsy. *Electroencephalography and Clinical Neurophysiology.* 94:326-337.

[106] Darcey TM, Williamson PD. 1985. Spatio-temporal EEG measures and their application to human intracranially recorded epileptic seizures. *Electroencephalography and Clinical Neurophysiology.* 61:573-587.

[107] Gotman J, Levtova V, Farine B. 1993. Graphic representation of the EEG during epileptic seizures. *Electroencephalography and Clinical Neurophysiology.* 87:206-214.

[108] Hilfiker P, Egli M. 1992. Detection and evolution of rhythmic components in ictal EEG using short segment spectra and discriminant analysis. *Electroencephalography and Clinical Neurophysiology.* 82:255-265.

[109] Lehmann D, Michel CM. 1990. Intracerebral dipole source localization for FFT power maps. *Electroencephalography and Clinical Neurophysiology.* 76:271-276.

[110] Franaszczuk PJ, Bergey GK, Durka PJ, Eisenberg HM. 1998. Time-frequency analysis using the matching pursuit algorithm applied to seizures originating from the mesial temporal lobe. *Electroencephalography and Clinical Neurophysiology.* 106:513-521.

[111] Osorio I, Frei MG, Wilkinson SB. 1998. Real-time automated detection and quantitative analysis of seizures and short term prediction of clinical onset. *Epilepsia.* 39:615-627.

[112] Gath I, Lehmann D, Bar-On E. 1983. Fuzzy clustering of EEG signals and vigilance performance. *International Journal of Neuroscience.* 20:303-312.

[113] Gath I, Michaeli A, Feuerstein C. 1991. A model for dual channel segmentation of the EEG signal. *Biological Cybernetics.* 64:225-230.

[114] Lopes da Silva FH, Mars NJI. 1987. Parametric methods in EEG analysis. In: Gevins AS, Rémond A, editors. Methods of analysis of brain electrical and magnetic signals, *Handbook of Electroencephalography and Clinical Neurophysiology.* Amsterdam: Elsevier. p. 243-260.

[115] Schaller B, Ruegg SJ. 2003. Brain tumor and seizures: pathophysiology and its implications for treatment revisited. *Epilepsia.* 44(9):1223-32.

[116] Sanderson AC, Segen J, Richey E. 1980. Hierarchical modelling of EEG signals. *IEEE Transactions on Pattern Analysis and Machine Intelligence Intel.* 2:405-415.

[117] Wendling F, Badier JM, Chauvel P, Coatrieux JL. 1997. A method to quantify invariant information in depth-recorded epileptic seizures. *Electroencephalography and Clinical Neurophysiology.* 102:472-485.

[118] Wu L, Gotman J. 1998. Segmentation and classification of EEG during epileptic seizures. *Electroencephalography and Clinical Neurophysiology.* 106:344-356.

[119] Lehmann D, Ozaki H, Pal I. 1987. EEG alpha map series: brain micro-states by space-oriented adaptive segmentation. *Electroencephalography and Clinical Neurophysiology.* 67:271-288.

[120] Brekelmans GJF, van Emde Boas W, Velis DN, van Huffelen AC, Debets RMC, van Veelen CWM. 1995. Mesial temporal versus neocortical temporal lobe seizures: demonstration of different electroencephalographic spreading patterns by combined use of subdural and intracerebral electrodes. *J. Epilepsy.* 8:309-320.

[121] Wennberg R, Arruda F, Quesney LF, Olivier A. 2002. Preeminence of extrahippocampal structures in the generation of mesial temporal seizures: evidence from human depth electrode recordings. *Epilepsia.* 43(7):16-26.

[122] Bartolomei F, Wendling F, Bellanger JJ, Régis J, Chauvel P. 2001. Neural networks involving the medial temporal structures in temporal lobe epilepsy. *Clinical Neurophysiology.* 112:1746-1760.

[123] Bartolomei F, Wendling F, Vignal JP, Kochen S, Bellanger JJ, Badier JM, R B-J, Chauvel P. 1999. Seizures of temporal lobe epilepsy: identification of subtypes by coherence analysis using stereo-electro-encephalography. *Clinical Neurophysiology.* 110:1741-1754.

[124] Quesney LF, Risinger MW, Shewmon DA. 1993. Extracranial EEG evaluation. In: Engel JJ, editor. *Surgical treatment of the epilepsies.* New York: Raven Press. p. 173-195.

[125] Ebersole JS, Pacia SV. 1996. Localization of temporal lobe foci by ictal EEG patterns. *Epilepsia.* 37:386-399.

[126] Ebner A, Hoppe M. 1995. Noninvasive electroencephalography and mesial temporal sclerosis. *Journal of Clinical Neurophysiology.* 12(1):23-31.

[127] Foldvary N, Klem G, Hammel J, Bingaman W, Najm I, Luders H. 2001. The localizing value of ictal EEG in focal epilepsy. *Neurology.* 57(11):2022-2028.

[128] Risinger MW, Engel JJ, Van Ness PC, Henry TR, Crandall PH. 1989. Ictal localization of temporal lobe seizures with scalp/sphenoidal recordings. *Neurology.* 39:1288-1293.

[129] Williamson PD, French JA, Thadani VM, Kim JH, Novelly RA, Spencer SS, Spencer DD, Mattson RH. 1993. Characteristics of medial temporal lobe epilepsy: II. Interictal and ictal scalp electroencephalography, neuropsychological testing, neuroimaging, surgical results, and pathology. *Annals of Neurology.* 34(6):781-787.

[130] Foldvary N, Lee N, Thwaites G, Mascha E, Hammel J, Kim H, Friedman AH, Radtke RA. 1997. Clinical and electrographic manifestations of lesional neocortical temporal lobe epilepsy. *Neurology.* 49(3):757-763.

[131] O'Brien TJ, Kilpatrick C, Murrie V, Vogrin S, Morris K, Cook MJ. 1996. Temporal lobe epilepsy caused by mesial temporal sclerosis and temporal neocortical lesions. A clinical and electroencephalographic study of 46 pathologically proven cases. *Brain.* 119:2133-2141.

[132] Blanke O, Lantz G, Seeck M, Spinelli L, Grave de Peralta R, Thut G, Landis T, Michel CM. 2000. Temporal and spatial determination of EEG-seizure onset in the frequency domain. *Clinical Neurophysiology.* 111:763-772.

[133] Lantz G, Michel CM, Seeck M, Blanke O, Landis T, Rosén I. 1999. Frequency domain EEG source localization of ictal epileptiform activity in patients with partial complex epilepsy of temporal lobe origin. *Electroencephalography and Clinical Neurophysiology.* 110:176-184.

[134] Jung WY, Pacia SV, Devinsky R. 1999. Neocortical temporal lobe epilepsy: intracranial EEG features and surgical outcome. *Journal of Clinical Neurophysiology.* 16(5):419-425.

[135] Schwartz BE, Halgren E, Delgado-Escueta AV, Feldstein P, Maldonado H, Walsh GO. 1990. Multidisciplinary analysis of patients with extratemporal complex partial seizures. II. Predictive value of semeiology. *Epilepsy Research.* 5:146-154.

[136] Bancaud J. 1969. Les crises épileptiques d'origine occipitale (etude stereoélectroencephalographique). *Revue Otoneuroophtalmologie.* 41:299-314.

[137] Williamson PD, Spencer S. 1986. Clinical EEG features of complex partial seizures of extratemporal origin. *Epilepsia.* 27 (suppl 2):S46-63.

[138] Quesney LF, Gloor P. 1985. Localization of epileptic foci. In: Gotman J, Ives JR, Gloor P, editors. Long-term monitoring in epilepsy. Amsterdam: Elsevier Science Publishers.

[139] Salanova V, Andermann F, Rasmussen T, Olivier A, Quesney LF. 1995. Parietal lobe epilepsy. Clinical manifestations and outcome in 82 patients treated surgically between 1929 and 1988. *Brain.* 118:607-627.

[140] Williamson PD, Boon PA, Thadani VM, Darcey TM, Spencer DD, Spencer SS, Novelly RA, Mattson RH. 1992a. Parietal lobe epilepsy: diagnostic considerations and results of surgery. *Annals of Neurology.* 31(2):193-201.

[141] Williamson PD, Thadani VM, Darcey TM, Spencer DD, Spencer SS, Mattson RH. 1992b. Occipital lobe epilepsy: clinical characteristics, seizure spread patterns, and results of surgery. *Annals of Neurology.* 31(1):3-13.

[142] Holmes MD, Brown M, Tucker DM. 2004. Are "generalized" seizures truly generalized? Evidence of localized mesial frontal and frontopolar discharges in absence. *Epilepsia.* 45(12):1568-1579.
[143] van Boguert P. 2005. Imaging of epilepsia in animals-PET and autoradiography. *Neuroscience Imaging.* 1: in press.
[144] Bernasconi A. 2005. Magnetic resonance spectroscopy in animal models of epilepsy. *Neuroscience Imaging.* 1: in press.
[144] Schaller B. 2005. Influences of brain tumor-associated pH changes and hypoxia on epileptogenesis. *Acta Neurol. Scand.* 111(2):75-83.

*Chapter 5*

# CUSTOMIZED TISSUE ENGINEERING IN HEAD AND NECK

## *R. Staudenmaier[1]\*, J. Welter[3], G. Meier[2] and M.M. Wenzel[1]*

[1] Department of ENT, Head and Neck Surgery, University Hospital,
Technical University Munich, Germany.
[2] PolyMaterials, Kaufbeuren, Germany.
[3] KL-Technik, Gauting, Germany.

## ABSTRACT

Tissue Engineering (TE) of cartilage for reconstructive surgery seems to be a promising option to gain tissue for three-dimensional (3D) structures with a minimal donor site morbidity. A specific defect needs a customized implant. Most TE strategies rely on the application of resorbable three-dimensional (3D) scaffolds to guide the growing tissue. Each tissue requires a specific scaffold with precisely defined properties which depend on physiological environment.

Rapid Prototyping (RP) technologies allow to fabricate scaffolds with any geometric complexity, from multiple materials, as composites and even the inner architecture of the object can be varied in a defined manner and at a defined spot. Scaffolds can be manufactured by RP techniques directly from computer added design (CAD) data sources e.g. via STL file.

The combination of Tissue Engineering and Rapid Prototyping presents the basis for the realization of customized implants. It provides a helpful support for conventional ear reconstruction and gives new perspectives for the reconstructive surgery.

**Keywords:** Rapid Prototyping; Tissue Engineering.

---

\* Corresponding author: Staudenmaier Rainer, M.D.; Department of ENT and Head and Neck Surgery; University Hospital of the Technical University; Ismaninger Str. 22; 81675 München, Germany; Rainer.Staudenmaier@lrz.tum.de

## INTRODUCTION

In Head and Neck Surgery a variety of defects, iatrogenic, traumatic or inborn have to be augmented with supportive tissue. The ideal material is autogenous tissue, which is associated with limited resources and consecutive donor-site morbidity. Also the transplants often lack in dimension, shape and function. Tissue Engineering (TE) seems to be a promising alternative option to overcome these limitations. Most TE strategies rely on the application of resorbable three-dimensional (3D) scaffolds to guide the growing tissue. Different Rapid Prototyping (RP) technologies like stereolithography, solid freeform, fused deposition modelling, 3D-printing or negative molding offers the possibility to fabricate 3D models [1]. With slice data acquisition through computer or magnetic resonance tomography a defect is detectable in a mathematically exact way. Data processing enables computer added design (CAD) and computer added manufacturing (CAM). This technique is in clinical use for alloplastic materials for example as calivarium titanium implants [2]. It allows generating the ideal implant for an individual defect (figure 1).

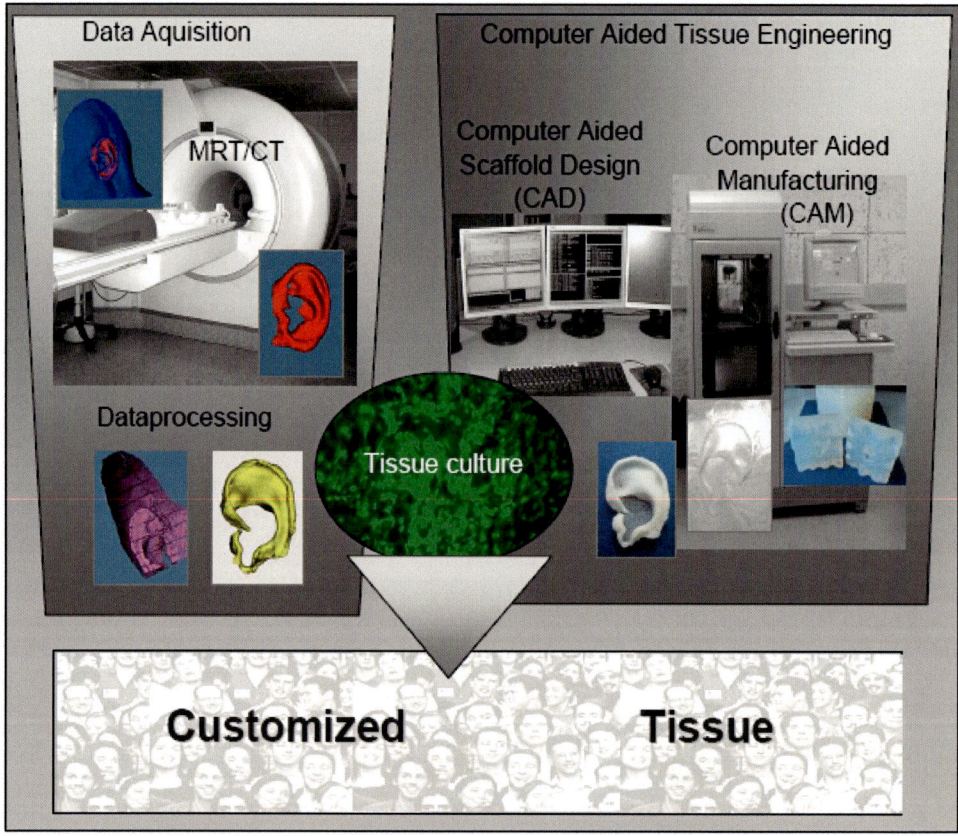

Figure 1. Overview of the way from data acquisition to a customized autologous implant.

Using this advantage generating scaffolds for TE with the help of RP is challenging [3], [4], [5], [6], [7]. Necessary scaffold structural features are high porosity, pore size, interconnectivity for cell seeding and nutrient diffusion. The material has to be non cytotoxic, should enable cell adhesion and proliferation. It has to provide specific biomechanical stability, elasticity, controllable degradation and resorption rates to match tissue replacement.

The so customized scaffold could be seeded with cells of various sources; with differentiated cells like chondrocytes or periostal cells for cartilage tissue, präadipocytes for fat tissue, mesenchymal stem cells for bone, cartilage or tendon TE. Every type of cells requires optimal material characteristic and seeding strategies to generate the desired tissue. For optimal tissue development specific supplements and growth factors are essential. Also a variety of bioreactors are in use to support tissue development and tissue architecture for clinical application.

This article should give a survey of the combination of RP and TE for generating individual autologous implants with emphasis on potential clinical applications in Head and Neck Surgery.

## IMAGING DATA ACQUISION

In order to receive a scaffold finally the basis of the manufacturing process – the imaging data acquisition – has to be appropriate. That means the resolution of the slice images has to be high enough that all necessary details are represented. But it has to be taken into account that even with today's fast computers very huge datasets are hard to handle. In principle setting resolutions on image acquisition scanners like CT and MR that they do not natively support but that are computed by interpolation is contra productive as it harms data transmission and does not lead to better quality. Today very advanced cubical interpolation and filtering algorithms are available but details that are not contained in the basis image material cannot be received in the following data processing.

With reference to the used image acquisition technology the head start of CT-scanners in image resolution and quality is so huge that their application is adequate in most cases. This includes the representation of cartilage structures like stated later on.

The requirements for calvarium models and implants respectively are at least critical. Experience shows, that slice distances in the range between 1mm and 2mm combined with a planar pixel size of 0.5mm to 0.7mm are optimal for later reconstruction. Smaller values of these parameters do not lead to significantly better results but complicate data processing. With jaw joints and teeth a higher resolution of 0.5mm isotropic is needed to gain useable results. This also applies to complex cartilage structures like the concha. But in that special case additionally a high number of tonal values – at least 3000 unique ones - are needed to ensure a proper segmentation later on.

The image processing is done on personal computers. Because of this a data interchange format has to be chosen. Nowadays this can be done using DICOM format slice images on CD-R media in most cases. In order to ensure an unobstructed process images shall not be compressed. When using WORM and MO media it has to be considered, that this devices and the used formats are mostly proprietary to the manufacturer of the scanner. The usability of these devices has to be checked with the data processor.

## IMAGING-BASED 3D DATA PROCESSING

After importing the datasets on a PC workstation they can be processed with special software products that are commercially available. The most common are Mimics from Materialise (Leuven, Belgium), and Amira from Mercury Computer Systems (Chelmsford, USA). This software's goal is to generate a three-dimensional computer model on basis of the 2D slice images. Furthermore it provides supporting services like editing, measuring and direct interfacing to rapid prototyping machines. It has to be considered that in most cases multiple subdatasets cannot be combined at this point anymore. Therefore the desired regions have to be contained in one continuous record.

Different tissue types are represented in the scanner images by varying grey values wherefore the segmentation of the target type has to be made by choosing a certain bandwidth. This task tends to be intuitive as it depends on the calibration of the imaging equipment and reconstruction method. It is performed on all slices present with the same values automatically. On the basis of the resulting mask a cubical interpolation with a subsequent smoothing is done to receive a continuous virtual 3D model. This can be for example the bone structure of the head or all cartilage parts of it. In subsequent steps subsections can be separated by setting distraction layers.

With the mentioned proceeding a model of the existing structures is received. In order to get a scaffold for further processing the mockup can be used in different ways. In the case of the concha it simply can be mirrored because of its symmetry and small area of support. Implants for calvarium defects can often be obtained with the help of additional datasets made before excision. These two models can be mapped and subtracted that results in very accurate results in regard to fitting and appearance (figure 2).

If no guide information from the corresponding patient is available extrinsic ones can be used as a basis for constructing the implant piece by piece by means of standard CAD systems like used in engineering.

After having fully machined the computer model can be exported to one of many data formats used in rapid prototyping. Common ones are STL as a 3D exchange format with limited possible resolution and CLI which is already adapted to the RP-machines and cannot be edited afterwards therefore.

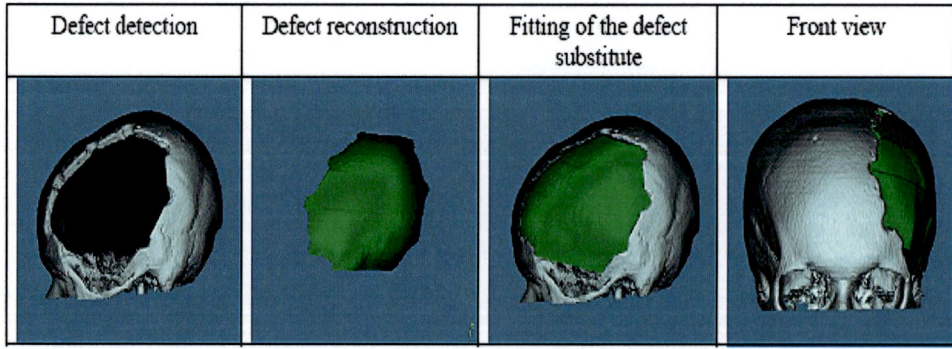

Figure 2. Virtual model of a calivarium defect.

# CUSTOMIZED SCAFFOLD DESIGN AND MANUFACTURING

Tissue engineering (TE) is a relatively young science which aims at producing tissue and organ substitutes. Because of limited resources of natural substitutes there is a huge worldwide need of artificial tissues and organs; as two big tissue groups - cartilage and bone. Considering cartilage defects, each year about 1 million patients are affected.

TE´s strategy is to combine patients own cells with specialized biomaterial scaffolds.

## Scaffolds

Scaffolds are three dimensional biomaterial structures which should mimic the function of the natural extracellular matrix within the tissue they should replace [9], [10]. Each tissue requires a specific scaffold with precisely defined properties. The scaffold design depends on physiological environment which is needed by the donor cells to build a proper tissue.

They serve as

- an adhesion substrat for the cell, fascilitating the localization an delivery of cells when they are implanted
- temporary mechanical support of the newly grown tissue by defining and maintaining a 3D structure
- guidance for the developed of new tissue with the appropriate function, e.g. by release of drugs like growth factors [11].

Scaffolds have to fulfill a set of requirements to be successfull in Tissue Engineering as biocompatibility, adapted surface- and physiochemical properties which promote or inhibit the initial cell attachment. Pore size, interconnectivity and pore morphology are responsible for the homogenous distribution of the cells into the scaffold and the trouble-free transport of nutrients, metabolites and degradation products which are the basis for tissue formation and vasculature [6], [8], [12], [13]. The mechanical properties have to be in line with the host environment, for example load-bearing tissue substitutes such as bone and cartilage [14], [15]. A sufficient mechanical strength and stiffness are required to deal with wound contraction forces and remodelling processes while tissue building and simultaneously scaffold degradation [12], [16]. At last an adjustable rate of degradation [17] – a smooth transition between tissue build-up and scaffold degradation, as well as a moderate immunological response to the scaffold material itself and the degradation products, are indispensable to be successful in building an efficient tissue substitute.

All kinds of different scaffold materials of natural and synthetical origin, degradable and non-degradable, have been investigated, e.g.:

- collagene, hyaluronic acid, fibrin, alginat and Pluronic for cartilage development [8]
- demineralized bone matrix, hydroxy apatite, ceramics, coral and glasses for bone [17], [18].
- Polylactic acid (PLA), (PGA) [19], polycaprolactone (PCL) [3] for bone and cartilage
- Poly(-ethylene glycol)-terphtalate PEGT for human nasal cartilage [13]
- Poly(L-lactide-co-glycolide) for nerve TE [20]

- Fibrin glue for cartilage [21] and urethral reconstruction [22].

An attempt to optimize the scaffold is represent by the fabrication of biomimetic polymers [23], [24], involvement of bioactive molecules (e.g. RGD sequenzes, heparan sulfate [25], [26] as cell linking spots [27]) or growth factors to enhance tissue formation. These active biomolecules can be dissolved or dispensed in the polymer or be chemically bonded to polymer backbone.

Furthermore scaffolds can be coated with proteins (by adhesion) like fibronectine and matrigel [28] to enhance cell connection to the surface.

Obviously the proper choice of an appropriate cell scaffold plays a fundamental role but is not simple [29].

Having detected the appropriate material for the wanted application, it must be checked if the material can be processed with the desired macro- and mirco-architecture and if the material is compatible with the thereto needed technical equipment. To meet the mentioned scaffold requirements TE uses Rapid Prototyping techniques [14], [30], [31], [32].

## Rapid Prototyping (RP) Technology

RP technologies are very specialized technologies known from industrial production of technical equipment [10].

RP technologies allow to fabricate physical objects directly from CAD data sources [33], [34], [35], [36], [37], [38] via STL file [39], [40], [41].

Objects can be formed with any geometric complexity without the need for final assembly. They can be fabricated from multiple materials, as composites and even the inner architecture of the object can be varied in a defined manner and at a defined spot. RP techniques can easily be automated and integrated with imaging techniques to produce scaffolds that are customized in size and shape [111], [112], [10]. In combination with TE the basis for customized implants is created.

There are more than 20 applied RP techniques in different application areas [12]. Solid free form fabrication (SFF) is an umbrella term for a number of different techniques such as Stereolithographie (SLA), Selective laser sintering (SLS), Three dimensional printing (3DP), 3D plotting and Fused deposition modelling (FDM). These SFF systems permit the fabrication of complex objects to a manageable, straightforward and relatively fast process.

Applying the know-how of RP on medical scaffold production, most of the macro- and micro-architectural requirements for TE applications can be satisfied [6]. Especially the microarchitecture is the precondition for the required discharge of nutrients and oxygen not only on the surface but also through the inlying areas of the scaffold [16].

However each of these RP technologies has its singular strengths and weaknesses [113] (Tab. 1).

All these RP techniques are highly specialized technologies concerning the material processability [114]. Not each scaffold material can be processed in each desired architecture with each RP equipment, what results in a significant reduction of the suitable biomaterials. That means the planning of a scaffold production has to consider if the wanted biomaterial is available in the required form such as solid pellet, powder, filament or solution. Next question must be if the characteristical material properties e.g. biocompatibility, are affected by the production process like solidification, heating etc.

It must be ensured that the choice of scaffold materials is compatible with the selected RP technique and that it is possible to fabricate the scaffold in the required way.

In the following only the techniques generally used for scaffold production in head and neck surgery will be described in short, advantages and disadvantages are named in table 1.

## Stereolithographie SLA, SLT

Stereolithographie is a photopolymerization technique used in manufacturing ridged parts. Objects are normaly made of photopolmer resins wich are hardened by the light from a laser beam. The laser is directed in x- and y-axis at the surface according to CAD data. A laser hardens a layer of liquid $\geq 0.5$ mm thick. The hardened layer is located on a stage which is lowered towards the z-axis, so that the solidified polymer is recovered with a new layer of polymer solution. In this way, layer by layer, the models cross-section is solidified and the shape from the CAD data is getting 3D reality. Support structures are needed to prohibit the loss of overhanging and unconnected components. To remove these support sturctures a manually model processing is nessesary.

Regarding the medical application, SLA can be used for the manufacturing of TE scaffolds, especially for surgery planning with the help of a 3D model and for building 3D models as educational tool for special surgeries. Furthermore the positive form can be fabricated to generate a mould – as for injection molding.

## Selective Laser Sintering SLS

Selective Laser Sintering (SLS) is a manufacturing technique that involves using a high temperature laser (for example, a carbon dioxide laser) to fuse polymer, ceramic or metal particle into desired shapes and layers. The laser rasters across a powder, elevates the powder temperature to reach the glass-transition temperature, causing the fusion of particles forming defined structures within each layer [12], [41]. The desired 3D object is build layer by layer. Complex porous ceramic scaffolds for bone tissue engineering present the typical application area of SLS [42], [43]. Resolutions of about 400-500µm can be achieved.

## Table 1. Comparison of different RP techniques used in TE scaffold fabrication for head and neck surgery

| RP technique | Resolution [μm] | Materials | Advantage | Disadvantage |
|---|---|---|---|---|
| Stereolithography SLA, SLT | 70-250 | Reactive resins, PEG acrylate, PEG methylacrylate, PVA, hyaluronic acid, dextran methacrylat, polypropylene fumarate | good mechanical strength; ease to use; easy to achieve small features | Limited to reactive resins (mostly toxic) must be photosensitive; extremly dense, low void volume |
| Selective Laser Sintering SLS | 400-500 | Bulk polymers, PE, ceramics, metals, compounds, PEEK, Polyvinylalkohol, PCL, PLLA, HA | high accuracy; high porosity; good mechanical strength; broad rande of bulk materials; no support structure needed; fast processing | Material must be in powder form; elevated temperatures – local high energy input; uncontrolled porosity; material shrinkage when sheet like structure is made; resolution depends on the laser beam diameter; powder may be trapped |
| 3D Printing | 100-500 | Ink + powder of bulk, Polymers, PLGA ceramics, starch, dextran, gelatine | no inherent toxic components; fast processing low costs; high porosity; can be performed in an ambient environement | weak bonding between powder particles; diminished accuracy; rough surface; component resolution and efficiency of removal of trapped materials is a concern |
| 3D Plotting | 100-250 | Swollen polymers (hydrogels), thermoplastic polymers, reactive resins, ceramics, PBT, PEOT | broad range of materials; broad range of conditions; Incorporation of cells, protein fillers | Slow processing; low accuracy; limited resolution; low mechanical strenght; no standard condition –time consuming adjustment to new materials |

| RP technique | Resolution [μm] | Materials | Advantage | Disadvantage |
|---|---|---|---|---|
| Fused Deposition Modelling FDM | 160-700 | Thermoplastic with good melt viscosities polymers/ceramics, PCL, PP-TCP, PCL-HA, PCL-TCP, PEGT-PBT | low costs; no trapped particles or solvents; highly reproducible; fully interconnected pore network; variation of pore morphology across scaffold realizable; input material in pellet form; preparation time is reduced | elevated temperatures; range of bulk materials; limited by melting point and processing conditions; no natural materials; medium accuracy; positive value for pore channels is applied; high temperature; rigide filament; pore heights is determined by size of polymer fiber; no incorporation of biomolecules |
| Injection Molding IM High pressure at hight temperature low pressure at room temperture | 135-500 | Hydrogele, PCL, polyester, collagen, ceramics, PLGA | Broad range of materials: At low pressure, room temperature IM high accuracy; complex shapes and defined wall thickness can be fabricated reproducible; can be automated | concerning compression molding: thermal degradation of polymers; no defined porosity and wall thickness, skin formation on polymer surface |

## Three Dimensional Printing 3DP

Three Dimensional Printing is a technique that uses a binder solution which is deposited onto a biomaterial powder bed using an ink jet printer. The binder material selectively joins particles according to CAD data. The powder bed including the biomaterial layer is then lowered so that the next powder layer can be spread and selectively joined with the binder. The whole process takes place at room temperature [12]. This layer-by-layer process is repeated until the model is completed. After a heat treatment the unbound powder is removed and the scaffold is ready. 3D Printing allows the fabrication of parts of any geometry, and out of any material, including ceramics [44], metals, polymers, cellulose [6] and composites can be realised in a short periode of time. Resolutions of about 100-500µm are processable [45], local variations of material composition, microstructure, and surface texture feasible [46], [14].

## 3D Plotting

This technique is based on dispensing a plotting material, mostly hydrogels [8], into a plotting medium to cause solidification of the material. Using a temperature controlled plotting medium, the solidification of the material during plotting into the medium can be caused by precipitation reaction, phase transition or chemical reaction. The plotting medium must be adjusted to the biomaterial so that gravity force is compensated by buoyancy otherwise gravity will cause complex structures collapse. As a result of the right plotting medium – biomaterial adjustment support structures are not needed. This effect can be increased by using a thixotrope plotting medium. Adjusting the medium to the plotting material is a very timeconsuming part of 3D Plotting.

Hydrogel structures with pore sizes of 200-400µm have been produced [8], [9].

## Fused Deposition Modelling

FDM is a solid-based rapid prototyping method that extrudes material, layer-by-layer, to build a model. Thermoplastic or ceramic filaments are melted inside a heated liquefier and then extruded through a nozzle, a specially designed head onto a platform to create a two-dimensional cross section of the desired model. The platform is lowered while the first layer is solidifiying. The next layer is extruded upon the previous layer [3]. This is repeated until the model is complete. A temperature-controlled environment is used to maintain sufficient fusion energy between each layer. Pore size depends on the dimension of the filaments extruded through the nozzle. Pore sizes of 160-700µm have been reported [14], [15], [47].

The range of suitable materials for FDM is limited because of the particular melting points, the needed good melt viscosities [12] and the processing conditions. Natural materials can´t be used because of the necessity to melt the material before extrusion. The need of high temperatures eliminates the incorporation of biomolecules. Therefore the possibility to produce a biomimetic surface by classical FDM technique is low.

Attempts to enlarge the range of suitable materials are presented by 3D fiber deposition technique, percision extrudering deposition PED and precise extrusion manufacturing PEM, low temperature deposition manufacturing, multinozzle deposition manufactoring, pressure assisted microsyringes and robocasting. Material melting is replaced by material dissolution. [10]. But even these promising techniques are connected with problems like need of solvents, high temperature, rigid filaments, low accuracy, low mechanical strength, low chance of incorporation of biomolecules.

## Injection Molding IM

Injection molding is a very widely used technique for manufacturing a variety of parts, from the smallest component to entire body panels of cars. Scaffolds are fabricated by using molds [7]. Molds can be fabricated according to a CAD data file by STL or 3D printing for example from metal, ceramic and silicone [48]. Liquid material is injected into a mold, which is the inverse of the desired shape [16], [48], [49].

In principle there are two different types of IM:

- molding with high compression and high temperature, liquified materials [13] and
- molding at low pressure and room temperature. [50]

Compression molding needs high pressure equipment, processing takes place at temperature [51] higher than the glass transition temperature of the materials. This implicates - in case of polymers - a thermal degradation to a certain degree. Scaffolds with small diameter and defined wall thickness are not easily to fabricate.

Low pressure molding takes place at room temperature. Thermal degradation of polymers does not appear.

Injection molding at low pressure, low temperature is characterized by high reproducibility, automatization and production efficiency. It enables the use of a lot of those biomaterials which are not compatible with the fabrication processing conditions of the above mentioned techniques. Complex shaped porous scaffolds but also tubular ones with thin walls and small diameter can be produced. Resolution of 135-500µm is feasible [50], [52], [53].

After the data acquisition, data processing and production of the desired scaffold with RP technology, it must be vitalised by cell seeding what represents the cellular part of tissue engineering.

## TISSUE ENGINEERING

According to Langer and Vacanti [54] Tissue engineering is defined as "an interdisciplinary field that applies the principles of engineering and the life sciences toward the development of biological substitutes that restore, maintain of improve tissue function".

The term 'tissue engineering' was officially minted at a National Science Foundation workshop (Whitaker Foundation ) in 1988 to mean the application of principles and methods of engineering and life sciences toward fundamental understanding of structure-function relationships in normal and pathological mammalian tissues and the development of biological substitutes to restore, maintain or improve tissue function.

Up to that time it was already known for years how to culture cells outside the body in two dimensional culture dishes. The innovation which comes along with TE as the first step towards to the development of three dimensional cell culture with the aim of producing complex three-dimensional tissue and organ substitues.

Urgent clinical needs are concerning skin, mucosa (e.g. burning or tumor victims), connective tissue, supporting tissue (cartilage, bone, joint, tendon, adipose tissue, muscle), peripheral nerves (e.g. for paraplegia), blood vessels, cornea, hole organs like liver, kidney, heart, heart valve as well as epithelial tissue for trachea and gastrointestinal tract [55], [56], [57], [58], [22], [59], [60], [61], [62], [63], [64].

In addition to the above named tissues some further applications which are in progress concerning head and neck surgery are auricle [65], [66], [1], laryngotracheal reconstruction [67], nasal tip [68] tympanic membrane [7] temporal bone [43], [39] calvarian defects [111], [48] vascularisation [69], [70] nerve, periphere and central NS [20], [71], [72], [73], [74], [75] mucosa [76], [77], [78], [79], [80], [81].

To realize these tissues TE´s strategy is the

- implantation of the isolated cells directly into the defect (cell-based therapie)
- implantation of a bioactive scaffold material which implements the ingrowth of the desired tissue (e.g. joint repair, implantation of scaffold material, stem cells shall migrate into the scaffold and rebuild bone or cartilage)
- implantation of a cell/scaffold combination

These attemps shall improve or replace biological functions. [82]. It sounds very easy but it is one of the most challenging plans in medical care.

## Cells

Which cells are used in TE?

- Differentiated autologous cells: highly specialised cells e.g. chondrocytes, keratinocytes. Autologe means that the cells come from the same body as that to which they will be reimplanted. Using autologous cells there is no risk of immunogenic rejection of the neotissue.
- Allogenic cells: allogenic means cells come from another body, e.g. stem cells
- Stem cells: autologe, allogenic (from bone marrow), progenitor cells (e.g. preadipocytes) and embryonal stem cells

*Differentiated autologous cells* are highly specialized cells characteristically for each single tissue with a clearly defined function, e.g. chondrocytes for cartilage. Using these cells for building a substitute means to reduce the limited mass of the donor tissue and to create therewith a second damage. A further serious problem is that the differentiated cells - taken out of the normal 3D environemt - tend to dedifferentiate in 2D cellculture environment within a few days. Dedifferentiation means, they are not producing tissue specific matrix, change their morphology and after a longer period of time even their genotype. To achieve the typical cell qualities again, cells have to be cultured in an appropriate 3D environment or scaffold.

Cells like cartilage cells, keratinocytes, muscle cells proliferate rapidly. Others like hepatocytes, cardiomyocytes proliferate slowly or even not [65]. Therefore alternative sources of cells - like stem cells - are needed.

*Stem cells* have the ability to transform into any other cell type and to divide without limit what is called "stem cell plasticity" [83]. In theory that means they have the potential to be the universal component to grow tissue substitutes of all kinds of tissues.

They are categorized in totipotent, pluripotent, multipotent and progenitor stem cells. Totipotent stem cells are cells which have the ability to grow into any other cell type. Pluripotent cells have the same possibility like totipotent stem cells. Multipotent stem cells can only grow into cells of a closely related family of cells. Progenitor cells can produce only one cell type (e.g. chondroblast turn into chondrocytes); but in contrast to non-stem cells, they have the property of self-renewal.

Furthermore a distinction is drawn between adult stem cells and embryonic stem cells.

Adult stem cells are undifferentiated cells found among differentiated cells of a specific tissue and are mostly multipotent cells [83].

Embryonic stem cells are obtained from the undifferentiated inner mass cells of an early stage embryo (bastocyst) that consists of 50 to 150 cells. They are thought to have much greater developmental potential than adult stem cells but their use is closely connected with ethical concerns [84].

Adult mesenchymal stem cells (MSC) can be obtained from blood from the placenta and umbilical cord remaining after birth (hematopoetic stem cells (MhSC)). One possible application is the cell-based therapie, that means the processed blood is injected into a vein of the patient which suffers from instance of acute lymphatic leukemia [85], Hurler syndrome, Hunter syndrom. For TE in combination with biomaterial scaffolds the isolated stem cells are expanded in vitro and used for the fabrication of tissue substitutes [86].

Further resources for adult stem cells are

- bone marrow, so called mesenchymal hematopoetic stem cells (MhSC); they are thought to be suitable for cartilage [87], [65], bone, liver, nerve, adipose tissue [88], [89], muscle, hair follicle, kidney substitutes [85], [90]
- fat; thought to be suitable for bone, fat, muscle, cartilage [91] and neurons. Fat stem cells seems to be similar to bone marrow stem cells, except that it is possible to isolate many more cells [92].
- peripheral blood [85], [93], [94]
- and in the organs themselfes.

Stem cells have also been identified in most tissues e.g. skin, blood vessels [95], dental pulp, digestive epithelium, retina, liver [96], brain [97], [98], [99].

Several types of heart diseases and for example myeloma have been examined using adult stem cells originating in bone marrow and peripheral blood [93], [94].

Using adult stem cells is a very important research domain because the availability of the cells is much higher than e.g. embryonal stem cells therefore adult stem cells can be used without ethical concerns.

## Cell Seeding

Having chosen the right cell type and cell source it must be decided how to get the cells into the scaffold.

Single Cells can be dropped, injected, sucked as a highly concentrated cell suspension onto and into a scaffold. Assuming that the force of gravitiy will lead the cells through the whole thickness of the scaffold. That strategy is often associated with cell loss. On the one hand they can fall through the pores into the culture dish and on the other hand they do not necessarily penetrate the scaffold. Reason for that can be, for example, an insufficient interconnectivity, or a hydrophobic surface.

Another possibility is the encapsulation of cells - macrocapsules, mircocapsules, 3D multicellular masses [100] – in hydrogels e.g. fibrin glue [22] alginate, Type I collagen, methylcellulose, pluronic F127 [19]. These particles can be seeded onto the material scaffold or the hydrogel/cell particles are shaped by a 3D plotter.

A very special kind of seeding is "seeding" by tissue induction. Thereto an unseeded bioactive scaffold is implanted which implements the ingrowth of the needed cells like stem cells from the bone marrow, which are migrating into the scaffold for knee joint repair.

Next step towards an implant is the cultivation of the vitalized biomaterials. For that purpose specialised bioreactors are build [101], [102], [103], [104] which enables to control the supply of nutrients, oxygen, the removal of metabolic waste and the stability of the pH-value. Some designs are equipped with cyclic, mechanical compression units to force the matrix development of chondrocytes [105].

Cell culture additives like growth factors are also a big concern in Tissue Engineering. Growth factors (GF) are defined as proteins that act as a signaling molecule between cells that attach to specific receptors on the surface of a target cell and promote differentiation and maturation of these cells. All kinds of GF´s are used in Tissue Engineering like Insulin-like-growth-factor (IGF), transforming-growth-factor beta (TGF-β), Interleukin (IL) for cartilage TE. Their use and benefit is described and discussed in a multitude of papers [106], [107], [108], [109], [110].

## Current Clinical Applications of TE Derived Products

Most of the current range of TE applications are experimental because it takes a long time until an TE product is developed and the medical implementation is achieved.

Tissue engineered bone is the only tissue type that has been evaluated in Phase III clinical trials and is in routine clinical use. [82] Engineered bone from BMP-2 and BMP-7 is used in orthopedics for lumbar fusions and long bone nonunions. Clinical application of bone tissue engineering in the head and neck is limited to case reports. There has been success in tissue engineering with cartilage for the nose and ear in immune incompetent animal models but there has been difficulty with generating scaffolds that do not incite an immune reaction in an immune competent model. The most advanced organ engineered is a bladder [115] that achieved success in a canine model.

## Variety of Tissue Engineering Biomedical Applications

http://www.collagenesis.com/documents/products.htm
Dermalogen® Human Tissue Matrix
Dermaplant™ Allograft Surgical Implant

http://www.collagenmatrix.com/technologyframe.html
Collagen matrices for several applications

http://www.genzymebiosurgery.com
EpicelTM cultured epidermal autografts - skin replacement
Carticel® autologous cultured chondrocytes - cartilage replacement

http://www.gensci.bc.ca/
DynaGraft® Matrix demineralized bone matrix

http://www.lifecell.com
AlloDerm® skin scaffold

http://www.millenium-biologix.com/
Skelite bone biomaterial

http://www.organogenesis.com/
Apligraf - tissue engineered skin
FortaFlex™ Family of Bioengineered Collagen Matrix Products

http://www.ortecinternational.com/
Composite Cultured Skin (OrCel™)

http://www.orquest.com
HealosTM mineralized collagen bone replacement

http://www.obi.com/
IMMIX™ cartilage repair products
IMMIX™ Extenders bone repair products

http://www.regenbio.com/
CMI Collagen Meniscus Implant

## Author's Own Application of Customized Tissue Engineering: Auricle for Ear Reconstruction

Our own field of interest is the total ear reconstruction. The first step is creating a 3D STL-file of the defect. Normal clinical used CT data can be used. They provide the possibility to separate cartilage from overlying soft tissue so that the actual cartilage defect can be detected. By means of Rapid Prototyping it is possible to generate models of the missing ear out of these data. These models can be used as templates for Tissue Engineering scaffolds. Since there had been no ideal biomaterial for Rapid Prototyping in Tissue Engineering we developed a totally new biomaterial which enables us to create an individual scaffold for Tissue Engineering of cartilage in means of reconstructive surgery. This 2-component material, based polycaprolacton, builts up an interconnective foam with a mechanical stability comparable to native cartilage. It shows homogenous 3D-structure, good cell adhaesion and biocompatibility. We evaluated first promising preliminary results for Tisssue Engineering of cartilage.

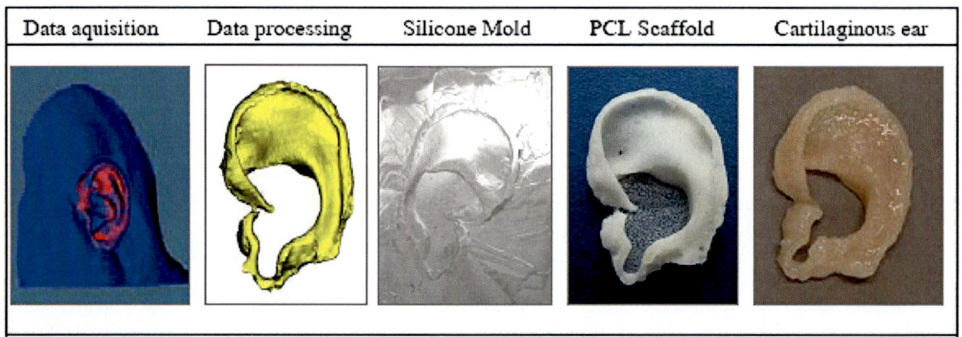

Figure 3. Auricle reconstruction by the means of Tissue Engineering; from data to scaffold and cartilaginous ear substitute.

*Future tasks for Tissue engineering* are an enhanced development of customized tissue engineering products to avoid the situation of missing donor tissue and unmeant immune response. Therefore the in-vitro culture of cells must be improved that cell dedifferentiation can be inhibited. Attempts to establish cocultures of different cells, the evolution of bioreactors which enables the cocultures of sized implants has to be forced. Computer-aided Tissue Engineering (CATE) will get more and more important to ensure the development of customized scafffolds with the appropriate technique, material, mirco- and macro-arcitecture without loss of geometrical resolution, accuracy in detail or material properties like biocompatibility.

## CONCLUSION

The combination of Tissue Engineering and Rapid Prototyping enables the next step in regenerative medicine: fabrication of customized implants. Customized implants are designed by the defect data gained of medical imaging technologies, transferd into the "speech" of RP machines, realised as biomaterial scaffold. These individual scaffolds can be seeded with autologous cells and after a short period of time implantation of the "own" implant will take place. Anyway the clinical application of customized implants as well as organ substitutes are still in infancy. Especially organ substitutes are not sufficient in first clinical trials, yet. Nevertheless the "TE-RP-anastomosis" represents a successful milestone on the way to remove the lack of donor tissue and donor organs.

## ACKNOWLEDGEMENTS

Our work was supported by Bayerische Forschungsstiftung, within the project FORTEPRO (Az.: 442/01) and international science exchange program (PIZ 17/03).

## REFERENCES

[1] Staudenmaier R, Naumann A, Bruning R, Englmeier KH, Aigner J. Ear reconstruction supported by a stereolithograpical model. *Plast. Rec. Surg.* 2000;106: 511-2.

[2] Dean D, Min KJ, Bond A. Computer aided design of large-format prefabricated cranial plates. *J. Craniofac. Surg.* 2003;14:819-32.

[3] Hutmacher DW, Schantz T, Zein I, Ng KW, Teoh SH, Tan KC. Mechanical Properties and cell cultural response of polycaprolactone scaffolds designed and fabricated via fused deposition modeling. *J. Biomed. Mater Res.* 2001;55:203-16.

[4] Landers R, Hubner U, Schmelzeisen R, Mulhaupt R. Rapid Prototyping of scaffolds from thermoreversible hydrogels and tailored for applications in tissue engineering. *Biomaterials.* 2002;23:4437-47.

[5] Taboas JM, Maddox RD, Krebsbach PH, Hollister SJ. Indirect solid free form fabrication of local and global porous biomimetic and composite 3D polymer-ceramic scaffolds. *Biomaterials.* 2003;24(1):181-94.

[6] Leong KF, Cheah CM, Chua CK. Solid freeform fabrication of three-dimensional scaffolds for engineering replacement tissues and organs. *Biomaterials.* 2003;24:2363-78.

[7] Hott ME, Megerian CA, Beane R, Bonassar LJ. Fabrication of tissue engineered tympanic membrane patches using computer-aided design and injection molding. *Laryngoscope.* 2004;114:1290-5.

[8] Landers R, Pfister A. Fabrication of soft tissue engineering scaffolds by means of rapid prototyping techniques. *J. Mater Sci.* 2002; 37:3107-3116.

[9] Moroni L, de Wijn JR, van Blitterswijk CA. Three-dimensional fiber-deposited PEOT/PBT copolymer scaffolds for tissue engineering: Influence of porosity, molecular network mesh size, and swelling in aqueous media on dynamic mechanical properties. *J. Biomed. Mater Res. A.* 2005 Aug 23, Epub ahead of print.

[10] Yeong WY, Chua CK, Leong KF, Chandraselaran M. Rapid prototyping in tissue engineering: challenges and potential. *Trends Biotechnol.* 2004;22:643-52.

[11] Jansen JA, Vehof JW, Ruhe PQ, Kroeze-Deutman H, Kuboki Y, Takita H, Hedberg EL, Mikos AG. Growth factor-loaded scaffolds for bone engineering. *J. Control Release.* 2005;101:127-36.

[12] Hutmacher DW, Sittinger M, Risbud MV. Scaffold-based tissue engineering: rationale for computer-aided design and solid free-form fabrication systems. *Trends Biotechnol.* 2004;22:354-62.

[13] Miot S, Woodfield T, Daniels AU, Suetterlin R, Peterschmitt I, Heberer M, van Blitterswijk CA, Riesle J, Martin I. Effects of scaffold composition and architecture on human nasal chondrocyte redifferentiation and cartilaginous matrix deposition. *Biomaterials.* 2005;26:2479-89.

[14] Sun W, Darling A, Starly B, Nam J. Computer-aided tissue engineering: overview, scope and challenges. *Biotechnol. Appl. Biochem.* 2004;39:29-47.
[15] Sun W, Starly B, Darling A, Gomez C. Computer-aided tissue engineering: application to biomimetic modelling and design of tissue scaffolds. *Biotechnol. Appl. Biochem.* 2004;39:49-58.
[16] Hollister SJ. Porous scaffold design for tissue engineering. *Nat. Mater.* 2005;4:518-24.
[17] Vats A, Tolley NS, Polak JM, Gough JE. Gene expression: a review of clinical applications in otorhinolaryngology-head and neck surgery. *Clin. Otolaryngol. Allied Sci.* 2002;27:291-5.
[18] Karageorgiou V, Kaplan D. Porosity of 3D biomaterial scaffolds and osteogenesis. *Biomaterials.* 2005;26:5474-91.
[19] Terada S, Yoshimoto H, Fuchs JR, Sto M, Pomerantseva L, Selig MK, Hannouche D, Vacanti JP. Hydrogel optimization for cultured elastic chondrocytes seeded onto a polyglycolic acid scaffold. *J. Biomed. Mater Res.* A 2005 Sep 1; [Epub ahead of print]
[20] B BT, Gao W, Wang S, Ramakrishna S. Development of fibrous biodegradable polymer conduits for guided nerve regeneration. *J. Mater Sci. Mater Med.* 2005;16:367-75.
[21] Park SH, Park SR, Chung SI, Pai KS, Min BH. Tissue-engineered Cartilage Using Fibrin/Hyaluronan Composite Gel and Its In Vivo Implantation. *Artif. Organs.* 2005;29:838-45.
[22] Bach AD, Bannasch H, Galla TJ, Bittner KM, Stark GB. Fibrin glue as matrix for cultured autologous urothelial cells in urethral reconstruction. *Tissue Eng.* 2001;7:45-53.
[23] Drotleff S, Lungwitz U, Breunig M, Dennis A, Blunk T, Tessmar J, Gopferich A. Biomimetic polymers in pharmaceutical and biomedical sciences. *Eur. J. Pharm. Biopharm.* 2004;58:385-407.
[24] Tessmar J, Kellner K, Schulz MB, Blunk T, Gopferich A. Toward the development of biomimetic polymers by protein immobilization: PEGylation of insulin as a model reaction. *Tissue Eng.* 2004;10:441-53.
[25] Mapili G, Lu Y, Chen S, Roy . Laser-layered microfabrication of spatially patterned functionalized tissue-engineering scaffolds. *J. Biomed. Mater Res. B. Appl. Biomater.* 2005 Jul. 15, Epub ahead of print.
[26] Hersel U, Dahmen C, Kessler H. RGD modified polymers: biomaterials for stimulated cell adhesion and beyond. *Biomaterials.* 2003;24:4385-415.
[27] Wilson CJ, Clegg RE, Leavesley ID, Pearcy MJ. Mediation of biomaterial-cell interactions by adsorbed proteins: a review.*Tissue Eng.* 2005;11:1-18.
[28] Kleinmann HK, McGarvey ML, Hassell JR, Star VL, Cannon FB, Laurie GW, Martin GR. Basement membrane complexes with biological activity. *Biochemistry.* 1986;25:312-8.
[29] Rice MA, Dodson BT, Arthur JA, Anseth KS. Cell-based therapies and tissue engineering. *Otolaryngol. Clin. North Am.* 2005; 38:199-214
[30] Sun W, Lal P. Recent development on computer aided tissue engineering--a review. *Comput. Methods Programs Biomed.* 2002;67:85-103.
[31] Webb PA. A review of rapid prototyping (RP) techniques in the medical and biomedical sector. *J. Med. Eng. Technol.* 2000;24:149-53.

[32] Tsang VL, Bhatia SN. Three-dimensional tissue fabrication. *Adv. Drug Deliv. Rev.* 2004;56:1635-47.
[33] Cheah CM, Chua CK, Leong KF, Cheong CH, Naing MW. Automatic algorithm for generating complex polyhedral scaffold structures for tissue engineering. *Tissue Eng.* 2004;10:595-610.
[34] Eufinger H, Wehmoller M, Machtens E, Heuser L, Harders A, Kruse D. Reconstruction of craniofacial bone defects with individual alloplastic implants based on CAD/CAM-manipulated CT-data. *J. Craniomaxillofac. Surg.* 1995;23:175-81.
[35] Hoffmann B, Sepehrnia A. Taylored implants for alloplastic cranioplasty - - clinical and surgical considerations. *Acta Neurochir. Suppl.* 2005; 93:127-9.
[36] Wallner CP, Roehrer-Ertl O, Schneider K. State-of-the-art computed tomography of primate skulls-comparison of different scan-protocols. *Ann. Anat.* 2004; 186:521-4
[37] Semple JL, Woolridge N, Lumsden CJ. In vitro, in vivo, in silico: computational systems in tissue engineering and regenerative medicine. *Tissue Eng.* 2005;11:341-56.
[38] Quadrani P, Pasini A, Mattiolli-Belmonte M, Zannoni C, Tampiere A, Landi E, Giantomassi F, Natali Casali F, Biagini G, Tomei-Minardi A. High-resolution 3D scaffold model for engineered tissue fabrication using a rapid prototyping technique. *Med. Biol. Eng. Comput.* 2005;43:196-9.
[39] Vorwerk U, Begall K. Practice surgery on the artificial temporal bone. Development of temporal bone facsimiles with stereolithography. *HNO.* 1998;46:246-51.
[40] Ciocca L, Scotti R. CAD-CAM generated ear cast by means of a laser scanner and rapid prototyping machine. *J. Prosthet. Dent.* 2004;92:591-5.
[41] Winder J, Bibb R. Medical rapid prototyping technologies: state of the art and current limitations for application in oral and maxillofacial surgery. *J. Oral Maxillofac. Surg.* 2005;63:1006-15.
[42] Tan KH, Chua CK, Leong KF, Cheah CM, Gui WS, Tan WS, Wiria FE. Selective laser sintering of biocompatible polymers for applications in tissue engineering. *Biomed. Mater Eng.* 2005;15:113-24.
[43] Suzuki M, Ogawa Y, Kawano A, Hagiwara A, Yamaguchi H, Ono H. Rapid prototyping of temporal bone for surgical training and medical education. *Acta Otolaryngol.* 2004;124:400-2.
[44] Seitz H, Rieder W, Irsen S, Leukers B, Tille C. Three-dimensional printing of porous ceramic scaffolds for bone tissue engineering. *J. Biomed. Mater Res. B. Appl. Biomater.* 2005;74:782-8.
[45] Lee M, Dunn JC, Wu BM. Scaffold fabrication by indirect three-dimensional printing. *Biomaterials.* 2005;26:4281-9.
[46] Mironov V, Boland T, Trusk T, Forgacs G, Markwald RR. Organ printing: computer-aided jet-based 3D tissue engineering. *Trends Biotechnol.* 2003: 21:157-61.
[47] Zein I, Hutmacher DW, Tan KC, Teoh. Fused deposition modeling of novel scaffold architectures for tissue engineering applications. *Biomaterials.* 2002;23:1169-85.
[48] Chang S, Tobias G, Roy A, et al.: Tissue engineering of autologous cartilage for craniofacial reconstruction by injection molding. *Plast. Reconstr. Surg.* 2003;112:793-801.
[49] Gomes ME, Ribeiro AS, Malafaya PB, Reis RL, Cunha AM. A new approach based on injection moulding to produce biodegradable starch-based polymeric scaffolds: morphology, mechanical and degradation behaviour. *Biomaterials.* 2001;22:883-9.

[50] Wu L, Jing D, Ding J. A "room-temperature" injection molding/particulate leaching approach for fabrication of biodegradable three-dimensional porous scaffolds. *Biomaterials.* 2006;27:185-191.

[51] Hou Q, Grijpma DW, Feijen J. Porous polymeric structures for tissue engineering prepared by a coagulation, compression moulding and salt leaching technique. *Biomaterials.* 2003;24:1937-47.

[52] Sachlos E, Czernuszka JT. Making tissue engineering scaffolds work. Review: the application of solid freeform fabrication technology to the production of tissue engineering scaffolds. *Eur. Cell Mater.* 2003;5:29-39; *discussion.* 39-40.

[53] Sachlos E, Reis N, Ainsley C, Derby B, Czernuszka JT. Novel collagen scaffolds with predefined internal morphology made by solid freeform fabrication. *Biomaterials.* 2003;24:1487-97.

[54] Langer R., Vacanti J.P.Tissue engineering. *Science.* 1993;260:920-6.

[55] Arosarena O. Tissue engineering. *Curr. Opin. Otolaryngol. Head Neck. Surg.* 2005;13:233-41.

[56] Bucheler M, Haisch A. Tissue engineering in otorhinolaryngology. *DNA Cell Biol.* 2003;22:549-64.

[57] Bucheler M. Tissue engineering in otorhinolaryngology, head and neck surgery. *Laryngorhinootologie.* 2002;81 Suppl 1:S61-80.

[58] Shin H, Jo S, Mikos AG. Biomimetic materials for tissue engineering. *Biomaterials.* 2003;24:4353-64.

[59] Palminteri E, Lazzeri M, Guazzoni G, Turini D, Barbagli G. New 2-stage buccal mucosal graft urethroplasty. *J. Urol.* 2002;167:130-2.

[60] Schlote N, Wefer J, Sievert KD. Acellular matrix for functional reconstruction of the urogenital tract. Special form of "tissue engineering"? *Urologe. A.* 2004;43:1209-12.

[61] Garner JP. Tissue engineering in surgery. *Surgeon.* 2004;2:70-8.

[62] Cortesini R. Progress in tissue engineering and organogenesis in transplantation medicine. *Exp. Clin. Transplant.* 2003;1:102-11.

[63] Chan C, Berthiaume F, Nath BD, Tilles AW, Toner M, Yarmush ML. Hepatic tissue engineering for adjunct and temporary liver support: critical technologies. *Liver Transpl.* 2004;10:1331-42.

[64] Laurson J, Selden C, Hodgson HJ. Hepatocyte progenitors in man and in rodents--multiple pathways, multiple candidates. *Int. J. Exp. Pathol.* 2005;86:1-18.

[65] Shieh S, Terada S, Vacanti J: Tissue engineering auricular reconstruction: in vitro and in vivo studies. *Biomaterials.* 2004;25:1545-1557.

[66] Renner G, Lane RV. Auricular reconstruction: an update. *Curr. Opin. Otolaryngol. Head Neck. Surg.* 2004;12:277-80.

[67] Kamil SH, Eavey RD, Vacanti MP, Vacanti CA, Hartnick CJ. Tissue-engineered cartilage as a graft source for laryngotracheal reconstruction: a pig model. *Arch. Otolaryngol. Head Neck. Surg.* 2004;130:1048-51.

[68] Kamil SH, Kojima K, Vacanti MP, Bonassar LJ, Vacanti CA, Eavey RD. In vitro tissue engineering to generate a human-sized auricle and nasal tip. *Laryngoscope.* 2003;113:90-4.

[69] Kannan RY, Salacinski HJ, Sales K, Butler P, Seifalian AM. The roles of tissue engineering and vascularisation in the development of micro-vascular networks: a review. *Biomaterials.* 2005;26:1857-75.

[70] Xu J, Ge H, Zhou X, Yang D, Guo T, He J, Li Q, Hao Z. Tissue-Engineered Vessel Strengthens Quickly under Physiological Deformation: Application of a New Perfusion Bioreactor with Machine Vision. *J. Vasc. Res.* 2005;42:503-508.

[71] Evans GR. Approaches to tissue engineered peripheral nerve. *Clin. Plast. Surg.* 2003;30:559-63.

[72] Taras JS, Nanavati V, Steelman P. Nerve conduits. *J. Hand Ther.* 2005;18(2):191-7.

[73] Hou S, Xu Q, Tian W, Cui F, Cai Q, Ma J, Lee IS. The repair of brain lesion by implantation of hyaluronic acid hydrogels modified with laminin. *J. Neurosci. Methods.* 2005;148:60-70.

[74] Inada Y, Morimoto S, Moroi K, Endo K, Nakamura T. Surgical relief of causalgia with an artificial nerve guide tube: Successful surgical treatment of causalgia (Complex Regional Pain Syndrome Type II) by in situ tissue engineering with a polyglycolic acid-collagen tube. *Pain.* 2005;117:251-8.

[75] Zhang N, Yan H, Wen X. Tissue-engineering approaches for axonal guidance. *Brain Res. Rev.* 2005;49:48-64.

[76] Lauer G. Schimming R. Clinical application of tissue-engineered autologous oral mucosa transplants. *Mund. Kiefer. Gesichtschir.* 2002;6:379-93.

[77] Glod A, Sadjadi Z, Vanscheidt W. Autologe Keratinozyten und Meshgraft-Hauttransplantation als Kombinationstherapie bei großflächigen chronischen Ulcera – ein Fallbericht. *Ellipse.* 2001;17:55-57.

[78] Yokoo S, Tahara S, Tsuji Y, Hashikawa K, Hanagaki H, Furudoi S. Functional and aesthetic reconstruction of full-thickness cheek, oral commissure and vermilion. *J. Craniomaxillofac. Surg.* 2001;29:344-50.

[79] Butler CE, Navarro FA, Park CS, Orgill DP. Regeneration of neomucosa using cell-seeded collagen-GAG matrices in athymic mice. *Ann. Plast. Surg.* 2002;48:298-304. *Laryngol.* 2005;114:429-33.

[80] Kinoshita S, Koizumi N, Nakamura T. Transplantable cultivated mucosal epithelial sheet for ocular surface reconstruction. *Exp. Eye Res.* 2004;78:483-91.

[81] Izumi K, Feinberg SE, Terashi H, Marcelo CL, Evaluation of transplanted tissue-engineered oral mucosal equivalents in severe combined immunodeficient mice. *Tissue Eng.* 2003;9:163-74.

[82] Nussenbaum B, Teknos TN, Chepeha DB. Tissue engineering: the current status of this futuristic modality in head and neck reconstruction. *Curr. Opin. Otolaryngol. Head Neck. Surg.* 2004;12:311-5.

[83] Conrad C, Huss R. Adult Stem Cell Lines in Regenerative Medicine and Reconstructive Surgery. *J. Surg. Res.* 2005;124:201-8.

[84] Evans M. Embryonic stem cells: a perspective. Novartis Found Symp 2005;265:98-103; discussion. 103-6, 122-8.

[85] Lewis A. Autologous stem cells derived from the peripheral blood compared to standard bone marrow transplant; time to engraftment: a systematic review. *Int. J. Nurs. Stud.* 2005;42:589-96.

[86] Goessler UR, Hormann K, Riedel F. Tissue engineering with adult stem cells in reconstructive surgery (review). *Int. J. Mol. Med.* 2005;15:899-905.

[87] Hui JH, Ouyang HW, Hutmacher DW, Goh JC, Lee EH. Mesenchymal stem cells in musculoskeletal tissue engineering: a review of recent advances in National University of Singapore. *Ann. Acad. Med. Singapore.* 2005;34:206-12.

[88] Alhadlaq A, Tang M, Mao JJ. Engineered adipose tissue from human mesenchymal stem cells maintains predefined shape and dimension: implications in soft tissue augmentation and reconstruction. *Tissue Eng.* 2005;11:556-66.

[89] Gregors CA, Prockop DJ, Spees JL. Non-hematopoietic bone marrow stem cells: molecular control of expansion and differentiation. *Exp. Cell Res.* 2005;306:330-5.

[90] Caplan AI. Review: mesenchymal stem cells: cell-based reconstructive therapy in orthopedics. *Tissue Eng.* 2005;11:1198-211.

[91] Awad HA, Wickham MQ, Leddy HA, Gimble JM, Guilak F. Chondrogenic differentiation of adipose-derived adult stem cells in agarose, alginate, and gelatin scaffold. *Biomaterials.* 2004;3211-3222.

[92] Zuk PA, Zhu M, Mizuno H, Huang J, Futrell JW, Katz AJ, Benhaim P, Lorenz HP, Hedrick MH. Multilineage Cells from Human Adipose Tissue: Implications for Cell-Based Therapies. *Tissue Engineering.* 2001; 7: 211-228.

[93] Flores PC, Conte LG, Fardella BP, Araos HD, Alfaro LJ, Aravena RP, Gonzalez GN, Larrondo LM. Autologous transplant (AT) with peripheral-blood stem-cell rescue for multiple myeloma: A clinical experience. *Rev. Med. Chil.* 2005;133:887-93.

[94] Caplice NM, Doyle B. Vascular progenitor cells: origin and mechanisms of mobilization, differentiation, integration, and vasculogenesis. *Stem Cells Dev.* 2005;14:122-39.

[95] Humpert PM, Eichler H, Lammert A, Hammes HP, Nawroth PP, Bierhaus A. Adult vascular progenitor cells and tissue regeneration in metabolic syndrome. *Vasa.* 2005;34:73-8, 80.

[96] Bjerknes M, Cheng H. Gastrointestinal stem cells. II. Intestinal stem cells. *Am. J. Physiol. Gastrointest. Liver Physiol.* 2005;289:G381-7.

[97] Goldmann SA, Sim F. Neural progenitor cells of the adult brain. Novartis Found Symp. 2005;265:66-80; *discussion.* 82-97.

[98] Parenteau NL, Rosenber L, Hardin-Young J. The engineering of tissues using progenitor cells. *Curr. Top Dev. Biol.* 2004;64:101-39.

[99] Verfaillie CM. Multipotent adult progenitor cells: an update. Novartis Found Symp. 2005;265:55-61; *discussion.* 61-5, 92-7.

[100] Takezawa T. A strategy for the development of tissue engineering scaffolds that regulate cell behavior. *Biomaterials.* 2003;24:2267-75.

[101] Martin I, Wendt D, Heberer M. The role of bioreactors in tissue engineering. *Trends Biotechnol.* 2004;22:80-6.

[102] Raimondi MT, Boschetti F, Falcone L, Fiore GB, Remuzzi A, Marinoni E, Marazzi M, Pietrabissa R. Mechanobiology of engineered cartilage cultured under a quantified fluid-dynamic environment. *Biomech. Model Mechanobiol.* 2002;1:69-82.

[103] Vunjak-Novakovic G. The fundamentals of tissue engineering: scaffolds and bioreactors. Novartis Found Symp. 2003;249:34-46; *discussion.* 46-51, 170-4, 239-41.

[104] Detamore MS, Athanasiou KA. Use of a Rotating Bioreactor toward Tissue Engineering the Temporomandibular Joint Disc. *Tissue Eng.* 2005;11:1188-97.

[105] Angele P, Abke J, Kujat R, Faltermeier H, Schumann D, Nerlich M, Kinner B, Englert C, Ruszczak Z, Mehrl R, Mueller R. Influence of different collagen species on physicochemical properties of crosslinked collagen matrices. *Biomaterials.* 2004;25:2831-41.

[106] Richmond JD, Sage AB, Shelton E, Schumacher BL, Sh RL, Watson D. Effect of Growth Factors on Cell Proliferation, Matrix Deposition, and Morphology of Human

Nasal Septal Chondrocytes Cultured in Monolayer. *Laryngoscope.* 2005;115:1553-1560.

[107] Almarza AJ, Athanasiou KA. Evaluation of three growth factors in combinations of two for temporomandibular joint disc tissue engineering. *Arch. Oral. Biol.* 2005 Aug 13; [Epub ahead of print]

[108] Westerhuis RJ, van Bezooijen RL, Kloen P. Use of bone morphogenetic proteins in traumatology. *Injury.* 2005 Aug 24; [Epub ahead of print]

[109] Yamaguchi T, Sawa Y, Miyamoto Y, Takahashi T, Jau CC, Ahmet I, Nakamura T, Matsuda H. Therapeutic angiogenesis induced by injecting hepatocyte growth factor in ischemic canine hearts. *Surg. Today.* 2005;35:855-60.

[110] Granjeiro JM, Oliveira RC, Bustos-Valenzuela JC, Sogayar MC, Taga R. Bone morphogenetic proteins: from structure to clinical use. *Braz. J. Med. Biol. Res.* 2005;38:1463-73.

[111] Schantz T, Hutmacher DW, Lam CX, Brinkmann M, Wong KM, Lim TC, Chou N, Guldber RE, Teoh SH. Repair of calvarial defects with customised tissue-engineered bone grafts II. Evaluation of cellular efficiency and efficacy in vivo. *Tissue Eng.* 2003;9 Suppl. 1:127-39.

[112] Woodfield TB, Malda J, de Wijn J, Peters F, Riesle J, van Blitterswijk CA. Design of porous scaffolds for cartilage tissue engineering using a three-dimensional fiber-deposition technique. *Biomaterials.* 2004;25:4149-61.

[113] Yang S, Leong KF, Du Z, Chua CK. The design of scaffolds for use in tissue engineering. Part II. Rapid prototyping techniques. *Tissue Eng.* 2002;8:1-11.

[114] Yang S, Leong KF, Du Z, Chua CK. The design of scaffolds for use in tissue engineering. Part I. Traditional factors. *Tissue Eng.* 2001;7:679-89.

[115] Alberti C, Tizzani A, Piovano M, Greco A. What's in the pipeline about bladder reconstructive surgery? Some remarks on the state of the art. *Int. J. Artif. Organs.* 2004;27:737-43.

In: Neuroscience Imaging Research Trends
Editor: B. Schaller, pp. 93-108

ISSN: 978-1-60456-227-9
© 2008 Nova Science Publishers, Inc.

**Chapter 6**

# RECENT DEVELOPMENTS IN BRAIN IMAGING OF SCHIZOPHRENIA: A SELECTIVE REVIEW

*Vince Calhoun*[*,1,2,4] *and Godfrey D. Pearlson*[1,2]

[1] The MIND Institute, Albuquerque, New Mexico 87131
[2] Dept. of ECE, University of New Mexico, Albuquerque, New Mexico 87131
[3] Olin Neuropsychiatry Research Center, Institute of Living,
Hartford, Connecticut, 06106
[4] Dept. of Psychiatry, Yale University School of Medicine,
New Haven, Connecticut; 06520.

## ABSTRACT

The use of noninvasive brain imaging in schizophrenia has made significant advances in recent years. One key area which has grown significantly is the emphasis on the identification of image-based biological markers–which have the potential to impact the way that schizophrenia is diagnosed and treated. This is enhanced by the development of new image analysis techniques for identifying relevant imaging information and combining the information gleaned from the various imaging modalities. A third key factor here is the advent of multi-site imaging studies. Finally, a number of recent studies have examined the relationship between brain imaging and genetic data. Given the known importance of both genetics and environment in brain function, the integration of genetics with brain imaging has the potential to fundamentally change our understanding of human brain function in disease through the identification of imaging endophenotypes. In this review we discuss the current impact of these new areas and their future implications.

---

[*] Reprint Address (VC): Vince Calhoun, Ph.D., The MIND Institute; 1101 Yale Boulevard; Albuquerque, NM 87131; Tel: (505) 272-1817, Fax: (505) 272-8002; Email: vcalhoun@unm.edu

# INTRODUCTION

Schizophrenia is a complex condition with a diverse and heterogeneous clinical presentation, which makes it unlikely that the neural mechanism underlying the disorder is limited to a dysfunction of a circumscribed brain region. Several comprehensive recent reviews describe the brain regions and circuits implicated in various major mental illnesses and enumerate the affected areas[1-4]. Schizophrenia has been associated with both structural and functional abnormalities in neocortical networks including frontal, parietal, and temporal regions and is thought to involve a disturbance of coupling between large-scale cortical systems[5-7]{Pearlson, 1996 #16835}. This theory hypothesizes that some subtle anomalous brain development occurs in utero and manifests itself symptomatically years later during adolescence and early adulthood[8, 9]. The evidence supporting this hypothesis is circumstantial primarily because human brain development is not amenable to direct investigation. Nevertheless, subtle physical[10, 11] and cognitive[12] abnormalities present before the onset of schizophrenic illness, and abnormalities of cerebral structure reported in first episode patients[13], imply disturbed brain development. Moreover, neuropathological studies of frontal and temporal cerebral regions report abnormal cortical cytoarchitecture, with neurons being misplaced or mis-sized, which likely resulted during development[14, 15]. Such neurodevelopmental changes imply that cortical disconnection plays a role in the pathophysiology of schizophrenia[16].

Rather than our re-summarizing a catalog of brain regions involved in schizophrenia and related disorders, it is of greater intrinsic interest to provide a brief summary of some key findings, and also to review the contemporary themes that are serving to motivate structural and functional MRI brain imaging research in psychiatry. We now discuss four relevant areas including 1) the search for image-based biological markers of illness, ("biomarkers"), 2) the development of flexible analysis and data fusion methods, 3) the emergence of multi-site imaging studies, and 4) integration of genetics and imaging.

# IMAGE-BASED BIOMARKERS

Many mental illnesses, such as schizophrenia, bipolar disorder, depression, and others, currently lack definitive biological markers and their diagnosis relies primarily upon symptom assessment. A "biomarker" was defined by the 1998 NIH Biomarker Definitions Working Group as "a characteristic that is objectively measured and evaluated as an indicator of normal biologic processes, pathogenic processes, or pharmacologic responses to a therapeutic intervention". The concept of a biomarker subsumes that of an endophenotype, which is essentially a biomarker with a direct connection to a genetic condition, that thus is seen not only in individuals with manifest illness, but also occurs more commonly in persons at risk for the disease (e.g. unaffected close relatives) more commonly than in the general population. The study of brain imaging data for potential biomarkers—a measurable indicator of a disease process—has received much attention recently as a conceptual framework that may better reflect relationships between clinical phenotype and neurobiology. Mostly this work relies upon studying correlations between highly distilled measures derived from images of the brain (from, e.g. small regions of interest).

There has been some work already in developing biomarkers based upon structural imaging[17, 18], functional imaging[19], genetics[20], and neuropsychological scores, or a combination of multiple measures[21]. It should be noted that such a goal is not new, since it has long been thought that it might be possible to utilize biological information for diagnostic purposes. What is occurring more frequently is an emphasis upon using imaging data and upon combining imaging information with other types of data. In addition, the methods for incorporating imaging data, which are considerably more complex due to the large amount of data, are becoming more sophisticated.

Early studies necessarily focused upon image-based biomarkers derived from regions of interest from specific brain regions. The hypothesis that psychiatric disorders could arise from disruption, dysfunctional integration or "disconnection" either between regions within a distributed network of brain regions or between such neural circuits has gained traction, particularly for schizophrenia[22]. As a means of quantifying such disconnections, "functional connectivity," defined as correlation between two or more functional regions[23] or among brain regions[24-28] has considerable utility. For example, analyzing fMRI correlations between brain regions in schizophrenia patients, Liang, et al., reported disrupted functional integration of widespread brain areas[23].

In an example from our own work, we attempted to relate temporal lobe function to individual diagnostic status utilizing an intrinsic, task-uncorrelated measure. Using functional magnetic resonance imaging data, we calculated maps of temporally coherent activity using an advanced analysis approach called independent component analysis (ICA; discussed in more detail later). Patient maps revealed greater synchrony in anterior and lateral superior temporal gyrus regions; control maps had greater synchrony in posterior and medial regions (see Figure 1; for more details see [19]). A within-participant subtractive comparison of these two sets of right hemisphere temporal lobe regions (optimized for cohort 1 using a minimum probability of error criterion) differentiated schizophrenia from healthy controls with 97% accuracy initially and performed with 94% accuracy in a confirmatory study of new subjects scanned at a different geographic site. This approach and other such measures[17, 21], when evaluated in a population diverse enough to establish sensitivity and specificity to the disorder, have the potential to provide a powerful, quantitative clinical tool for the assessment of schizophrenia.

## IMAGE ANALYSIS TECHNIQUES

There have been some significant advances in image analysis techniques with application to the study of schizophrenia. We discuss two examples in this section, the first being the use of flexible analysis approaches which are capable of characterizing the complexity of the data. The second is the use and development of techniques for combining or fusing multiple imaging data types.

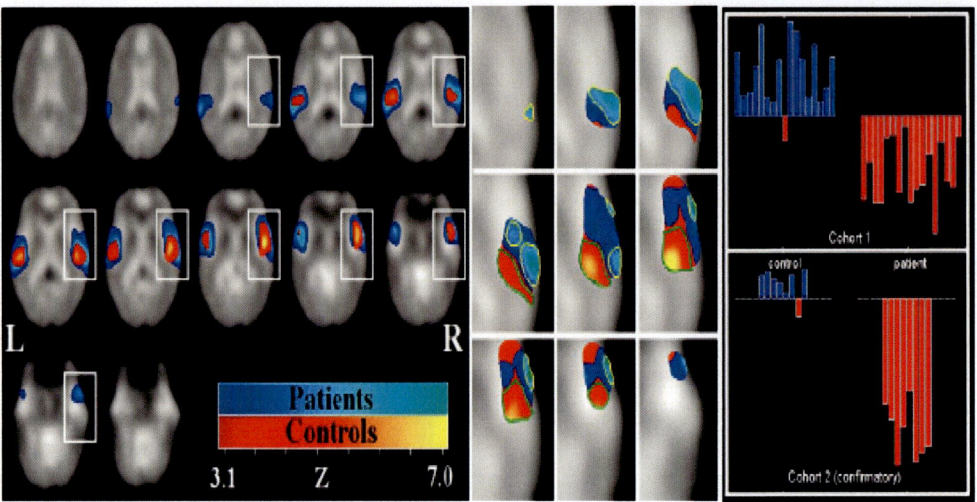

Figure 1. Classification of schizophrenia from healthy controls using fMRI data: (left) Mean activation maps from patients with schizophrenia and healthy controls. Right auditory cortex demonstrated the greatest difference (white box); (middle) Right auditory cortex difference maps with optimized boundaries; (right) Individual classification results for cohort 1, and replication in cohort 2. Schizophrenia classification is indicated with the color red.

## Flexible Analysis Approaches

The large amount of data potentially collectable in typical imaging studies, combined with a lack of understanding of brain impairment in mental illness has led to the emergence of more flexible analysis approaches in their ability to characterize brain function. A recent analytic approach, independent component analysis (ICA)[29-31], is a powerful means to decompose and understand complex data sets. ICA is a method for recovering underlying signals from linear mixtures of these signals and draws upon higher-order signal statistics to determine a set of "components" that are maximally independent of each other[32]. Spatial ICA is especially useful at revealing brain activity not anticipated by the investigator, and at separating brain activity into "networks" of brain regions with temporally synchronous BOLD fMRI signals. The primary assumption guiding spatial ICA, that "different parts of the brain do different things"[33] is consistent with the principle of modularity of brain function, and is used increasingly in fMRI data analysis.

Several recent studies have used ICA in three major types of circumstances. The first is to examine dominant activation differences in patients[31, 34], particularly during "resting state" scans, which are relatively easy to obtain and do not confound performance (e.g. difficulty or inability of patients to perform a task) with brain activity[35-37]. During the resting state, patients in the fMRI scanner are told to relax and not to focus their thoughts on any one particular topic. Such functional imaging studies, e.g. by Greicius et al. reveal that specific brain regions, including posterior cingulate cortex (PCC) and ventral anterior cingulate cortex (vACC), consistently show greater activity during resting states than when engaged in cognitive tasks[35, 37-39]. PCC is strongly coupled with vACC and several other brain regions at rest, and activity in these regions diminishes with the degree of engagement in cognitive tasks of increasing complexity, supporting the notion that these regions constitute a cohesive network supporting a default mode of brain function. A simple view is that resting state represents the brain in an "idling" mode. Thus, functional connectivity maps of PCC and vACC during a simple visual processing task are virtually identical to those obtained during rest[38]. However lateral prefrontal regions showing increased activity during a more complex cognitive task show significant inverse correlations among themselves and PCC, suggesting a mechanism for attenuation of default mode network activity during cognitive processing[38].

The resting state network has now been examined in several neuropsychiatric diseases. Greicius[37] examined an Alzheimer's disease (AD) group in the resting state and reported they displayed decreased resting-state activity in the posterior cingulate and hippocampus, compared to healthy controls. They suggested that that disrupted connectivity between these two regions accounts for the posterior cingulate hypometabolism commonly detected in positron emission tomography studies of early AD. A statistical analysis at the individual subject level suggested that activity in the default-mode network may ultimately prove a sensitive and specific biomarker for incipient AD[37]. Resting state fMRI scans are also abnormal in schizophrenia and autism[23, 36]. This has obvious repercussions for ultimate clinical diagnostic use; especially as resting state scans do not require cooperation with a cognitive task paradigm or any overt subject response.

## Multimodal Imaging and Analysis

Another area which has benefited from the development of new analysis tools in the fusion of data from different modalities (e.g. structural or function) or different function tasks. The brain is a complex organ of interrelated structural and functional connectivity and the importance of integrating anatomic information in functional studies is increasingly being recognized. A recent brain connectivity workshop concluded that "anatomical detail has an important role in the study of functional integration within the brain" .Passingham et al[40]., propose the concept of a "connectional fingerprint," i.e. that the functions of a cortical area are determined by its extrinsic connections and intrinsic properties . This suggests that there is an anatomical basis for functional connectivity, that anatomic information is a critical component of brain functional connectivity, and can be useful in distinguishing connection patterns. Due to the many available types of imaging methods, approaches, and models, it is important to define what is meant when discussing "brain connectivity". Relevant differences include spatial and temporal resolution, whether the data represent neuron activities, neural ensemble activities, or (electrical or hemodynamic) activities of macroscopic brain regions. As Horwitz et al.[41] point out, "using different definitions as to what constitutes functional connectivity may result in different conclusions about whether the associated functional connectivities for a particular neural system are strong or weak". For example, two widely used definitions in the fMRI community specify whether one is interested in 1) what regions are temporally similar (e.g. correlated), so called "functional connectivity", or 2) whether one region is influencing another region(s), called "effective connectivity". While Horwitz[41] argues for large-scale neural modeling to investigate relationships between various modalities and brain connectivity, he also points out that such analyses have many limitations. Connectivity is thus comprised of a class of concepts needing operational definition. A useful perspective is: the study of spatial and/or temporal correspondence in the brain. Achieving this requires 1) a correspondence metric (e.g. correlation, modulation) and 2) an imaging modality or modalities. The above definition subsumes both the concepts of functional[42, 43] and effective connectivity[44, 45], and also the concepts of individualized or population-based connectivity.

The study of schizophrenia can benefit greatly from the fusion of multimodal data, because the ability to examine the relationships between different brain modalities provides a powerful tool for better understanding this disorder. The imaging findings in schizophrenia are widespread, heterogeneous, not diagnostic and have limited replicability. It is likely that in part the lack of consistent findings is because most models do not combine modalities in an integrated manner and miss important changes which are only partially detected by different modalities individually. Combining modalities may thus uncover previously hidden relationships which can unify disparate findings in brain imaging of schizophrenia.

Data fusion as a technique is being newly utilized in brain imaging studies of various types. For example, it has been used to integrate temporal information in EEG data with spatial information in fMRI data[46]. Additional approaches to this type of problem involve constraining the EEG data with fMRI activation maps[47] or, constraining DTI fiber tracking using fMRI activation maps[42]. While these are powerful techniques, a limitation is that it may impose potentially unrealistic assumptions upon the DTI data, which are fundamentally of a different nature than the fMRI data. Additionally, such methods would not be able to detect a functional change associated with a distant white matter connection. More recent approaches have attempted to address this problem[48]. In addition, the use of flexible methods to examine joint associations between data types, for example examining brain connectivity using direct data fusion via a statistical framework (i.e. statistical data fusion) and source separation techniques is also useful for assessing relationships between distant regions[49]. We now present an example of such an approach in the context of two distinct functional networks.

A typical imaging experiment involves scanning participants during the performance of multiple types of images or multiple tasks. Because the brain is a highly interconnected organ, it is reasonable to expect that functional changes in one area may result in or be related to modulations of brain activity or brain structure in distant regions[50]. It may be useful to examine tasks (or conditions) which are theoretically related, but probe slightly different aspects of a particular functional domain (such as cognition or sensory processing). Such work may reveal, for example, different regions in the two tasks which are subsets of the main effect of each task but are activated by the tasks similarly. This would then provide new evidence of a link between these regions which could be further probed by additional experiments. Another approach is to use tasks which probe very different functional domains, each of which may be modulated by, *e.g.*, an underlying disorder such as schizophrenia. In such an approach, one might expect patients with schizophrenia to share a common, widespread deficit resulting in activation differences which transcend individual cognitive domains. This viewpoint is supported by the schizophrenia literature showing differences in almost every type of cognitive or sensory task studied[19, 51-54]. In this context, the use of tasks which probe different functional domains is advantageous, and joint analyses of these data may even help us to better unify the diversity of findings present in previous work.

Existing tools for examining joint information include region-based approaches such as structural equation modeling or dynamic causal modeling[55, 56]. For example, though it is most common to use, *e.g.*, structural equation modeling (SEM) to examining the relationship between different regions, these types of approaches can also be used to look at the correlational structure between regions activated by different tasks[57] or between functional and structural variables[58]. However these approaches do not provide an examination of the full set of brain voxels. A natural set of tools that avoid this problem include those that transform data matrices into a smaller set of modes or components. Such approaches include those based upon singular value decomposition[59, 60] as well as more recently, independent component analysis (ICA)[31].

In contrast to a first-level ICA approach (*e.g.* as discussed in the previous section), one can perform a second level (group), *feature-based* analysis of the fMRI activation maps (the 'features') generated from a first-level analysis[61]. In Figure 2, we show results from an analysis of fMRI data collected from two different tasks; auditory oddball[62] and Sternberg working memory[53, 63], both of which are known to be affected in patients with schizophrenia. Joint histograms show the activation levels for a joint functional network containing regions activated by either the auditory oddball or the Sternberg task.

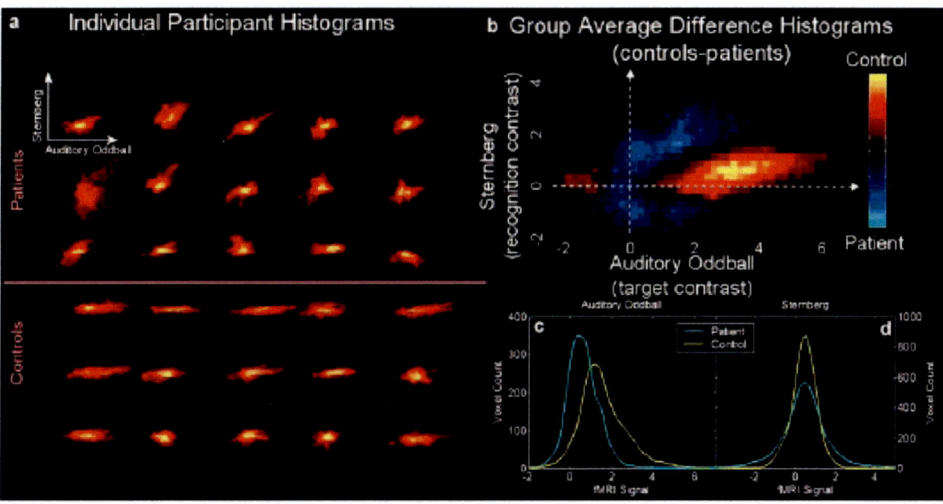

Figure 2. Integration of two different fMRI tasks: Joint 2D histograms for voxels identified in the jICA analysis. Individual (a) and group average difference (b) histograms [with orange areas larger in controls and blue areas larger in patients] are provided along with the marginal histograms for the auditory oddball (SPM contrast image for 'targets') (c) and Sternberg (SPM contrast image for 'recall') (d) data.

This approach revealed two important findings that were missed with traditional analyses. First, schizophrenia patients demonstrated overall "decreased" connectivity in the identified network. This network included regions in temporal lobe, cerebellum, thalamus, basal ganglia, and lateral frontal regions: these findings are consistent with both the cognitive dysmetria[64] and fronto-temporal disconnection[65] models. A second finding was that the correlation *between* the two tasks was significantly higher in patients than controls. This finding suggests that schizophrenia patients activate "more similarly" for both tasks than controls. The degree to which a brain activation map is different from that of another task may reflect the degree to which performance on a task is "specialized" to a certain set of regions. A possible synthesis of both findings is that patients are activating less, but also activating with a less unique set of regions for these very different tasks. This suggests both a global attenuation of activity as well as a breakdown of specialized wiring between cognitive domains. Alternative explanations are also possible, and methods such as we propose may prove useful for further study.

In summary, analysis methods are playing an increasingly important role in the use of imaging data. The use of flexible approaches as well as methods which are able to combine multiple image types into a unified analysis framework (to either provide a multi-dimensional view of brain function or to take advantage of the different strengths of imaging modalities such as EEG and fMRI) has significant advantages and is especially important for a multifaceted disease like schizophrenia.

## MULTI-SITE IMAGING STUDIES

The emergence of large-scale imaging studies is an important step, especially in establishing reliable biomarkers. A key reason for this is the need to increase the ability to identify intrinsically small differences in hard-to-recruit populations (such as prodromal subjects or patients with schizophrenia taking a particular medication). The hope is that by identifying subtle differences in these populations we may both learn something new about the disease and also define useful biological markers. Two fMRI initiatives which have some overlapping goals are the function BIRN (fBIRN) test bed-which is built upon a large-scale database infrastructure called the Biomedical Informatics Research Network (BIRN)-and the MIND Clinical Imaging Consortium/FIRST Program.

The major goal of the FBIRN (http://www.nbirn.net) is to develop tools to make multi-site functional MRI studies a common research practice. This is important since MRI scanners vary between site and between vendors, and characterizing, normalizing or removing the impact of these differences is necessary if the data are to be comparable between sites[66].

FBIRN also has a goal to test identified neural dysfunctions in schizophrenia. This is being tested using robust cognitive imaging tasks in a well-characterized, multi-site, representative sample of schizophrenia and control subjects. Representative samples are required to determine the generalizability of single site findings, especially in complex illnesses such as schizophrenia in which racial, gender, socioeconomic, and treatment setting differences can influence findings.

The MIND Institute (Mental Illness and Neuroscience Discovery, http://www.themindinstitute.org) was founded to advance the understanding of mental illness and mental disorders through the use of modern brain imaging techniques. It consists of neuroimaging scientists and medical research institutions including Harvard University, University of Minnesota, the University of New Mexico, and Los Alamos National Laboratory, and collaborates with other imaging centers such as The University of Iowa and the National Institutes of Health.

An essential part of developing useful image-based clinical tools is adequate patient assessment, and strengthening the link between the imaging data and the patient. MIND investigators have recently established the Functional Imaging Research for Schizophrenia Treatment (FIRST) clinics in order to provide expert assessment and treatment planning for patients at the outset of schizophrenic illness. These clinics will provide a platform for patients and control subjects for imaging, neuropsychology, and genetic testing. In addition the clinics will be able to provide subjects for pilot studies at individual sites. Patients will be sought through communication and advertisement in the region of each center by collaboration with psychiatrists, psychologists, primary care physicians, and schools. All subjects will receive a basic evaluation, including psychiatric interviews and laboratory testing. All subjects will be asked to participate in a research project and those consenting subjects will undergo imaging, neuropsychological testing, and a blood draw for DNA.

The fBIRN and MIND programs are just two examples of many other multisite initiatives. Another good example of multisite programs are those which study prodromes (those who exhibit early symptoms but may or may not eventually be diagnosed with schizophrenia)[67] or first episode patients[68] who are difficult to recruit in large numbers. For multi-site studies, current emphasis is focused upon the need to establish cross-site reliability. Once reliability is established, these and other multisite imaging programs will become a critical tool for reaching the long-term goal of image-based clinical tools.

## NEUROIMAGING AND GENETICS

Biomarkers may be primarily environmental, epigenetic, or multifactoral in their origin. For this reason, criteria useful for the identification of endophenotypes (defined earlier), a special subset of such markers for studies in psychiatric genetics have been proposed, adapted, and refined over time[69]. Current endophenotypic criteria are designed to direct clinical research in psychiatry toward genetically and biologically meaningful conclusions. "Imaging genetics" refers to the use of anatomical or physiological imaging technologies as phenotypic assays to evaluate genetic variation. Scientists that first explored imaging genetics were interested in how genes might influence psychopathology and used functional neuroimaging to investigate genes that are expressed in the brain[70].

Following rapidly on scientific efforts to unravel the human genome, advances in genetic knowledge have begun to exert an increasing influence on neuroimaging research. The functional mechanism of many of the disease 'risk genes' identified and the 'big picture' of how these genes combine in a common etiology or physiological final common path remains mostly obscure. However, it is becoming clearer that disease risk for these conditions is mediated by multiple, interacting genes of individually small effect, acting in concert, likely through convergent molecular "bottlenecks". With the faster paced, replicable identification of risk genes for schizophrenia and affective disorders that serve to motivate the field of psychiatric research, we have seen the recent emergence of more comprehensive, testable, etiopathologic models of major mental illnesses. Such models, advanced primarily by Weinberger and Murray[71-73] span genes, molecular and cellular biology, brain systems and behavior. These hypotheses have resulted in a series of interesting papers in which two or more of the above elements have been explored and related to one another.

Phenomenology-based diagnostic systems in psychiatry have several serious limitations resulting in classifications that are heterogeneous, lack clear boundaries, and match biological constructs poorly. The lack of clear diagnostic boundaries is particularly evident with respect to schizophrenia (SZ) and bipolar disorders where there is extensive overlap on many dimensions, including symptoms, neurophysiology, imaging, cognition, linkage "hotspots" and candidate genes, and treatment response. This has hindered progress in identifying vulnerability genes, clarifying etiopathophysiology and developing novel approaches to treatment. Phenotypic distinctions such as SZ or bipolar diagnoses have been widely held since Emil Kraepelin's time in late nineteenth century; however, when applied in linkage analyses, they have generally resulted in inconsistent findings, despite some interesting leads. Endophenotypes, defined more narrowly for our purposes as specific deficits in brain anatomy or function intermediate between the aberrant genes and the overt clinical disease, provide more direct clues to genetic underpinnings than the clinical syndrome[74]. The use of endophenotypes will help gene hunting in psychiatric illnesses, clarify diagnostic boundaries, and facilitate creation of relevant animal models.

As discussed earlier, the morphological and neuropathologic changes seen in schizophrenia suggest that schizophrenia is a disorder of connectivity. New insight has been provided into processes underlying this premise by the recent identification of several putative schizophrenia susceptibility genes. How might these genes relate to abnormal functional activation patterns in schizophrenia? Data from genetics, cognitive function and fMRI are beginning to converge. Recent studies are enlightening: Given notable working memory (WM) deficits in SZ, Egan[75] and Goldberg et al.[76] in a healthy subject cohort, examined a well-known functional val/met (v/m) catechol-omethyl–transferase (COMT) polymorphism in conjunction with an N-back working memory task[76]. There was a significant COMT genotype effect: v/v individuals had the lowest N-back performance and slowest reaction time and m/m individuals had the highest performance. Similar effects were also seen in patients with schizophrenia and their first-degree relatives. COMT genotype did not influence attention or IQ measures, suggesting specificity.

A sensible strategy, once a potential schizophrenia risk gene has been identified, is to explore whether or not normal SNP variants or 'flavours' of the gene have any influence on the normal brain structure or function. Discovery of how the gene impacts on the brain in this manner, particularly in disease-relevant regions, is important to document and helps provide a context as to how dysfunction at a genetic level plays out at a brain 'system' level. The DISC1 (disrupted in schizophrenia-1) gene is another promising schizophrenia candidate gene that is expressed predominantly within the hippocampus. Callicott[77] hypothesized that allelic variation at a Ser$^{704}$Cys SNP of DISC1 impacts hippocampal structure and function in healthy subjects. In fact, the Ser allele was associated with reduced hippocampal gray matter volume and altered activation of the hippocampus during several cognitive fMRI tasks. Together with other evidence suggesting that allelic variation within the DISC1 gene increases the risk for schizophrenia, one possible mechanism of this effect involves structural and functional alterations in the hippocampus.

In terms of impact on clinical practice, important insights into pathophysiology from imaging genomics will translate ultimately into both improved diagnosis, (based on genetic and imaging patterns alongside phenomenology). It may be years before these data are reflected in future versions of the DSM, but they will eventually be reflected in how we classify psychiatric illnesses. This in turn will lead to individually targeted treatments that address more specifically the varied pathophysiologic paths that can be reflected in similar phenomenologic states.

## CONCLUSION

Psychiatric research as a whole has been refreshed by the coalescence of ideas and findings from clinical and molecular genetics, cellular biology, brain structural and functional studies, engineering, statistics, and clinical phenomenology. The papers above are a small but representative selection from recent research. The success of the field depends upon the continued influx of scientists from different disciplines which is now occurring with increasing frequency. Scientists from specialties which only a few years ago were quite distinct from one another are now talking to each other increasingly and even having joint meetings, such as the recent one on genetics and neuroimaging.

## FUNDING SOURCES

VDC- 1 R01 EB 000840 NIBIB, 1 R01 EB 005846 NIBIB, NARSAD Young Investigator Award

GDP- 2 RO1 MH43775 NIMH MERIT Award, 5 RO1 MH52886, NIDA, 1 R01 DA020709, NIAAA, 1 RO1 AA015615, NARSAD Distinguished Investigator Award

## REFERENCES

[1] P. Fusar-Poli and M. R. Broome, "Conceptual issues in psychiatric neuroimaging.," *Curr. Opin. Psychiatry,* vol. 19, pp. 608-612, November 19 2006.

[2] M. T. Abou-Saleh, "Neuroimaging in psychiatry: an update," *Journal of Psychosomatic Research,* vol. 61, pp. 289-293, 2006/9 2006.

[3] M. T. Mitterschiffthaler, U. Ettinger, M. A. Mehta, D. Mataix-Cols, and S. C. R. Williams, "Applications of functional magnetic resonance imaging in psychiatry," *Journal of Magnetic Resonance Imaging,* vol. 23, pp. 851-861, 2006.

[4] H. Tost, G. Ende, M. Ruf, F. A. Henn, and A. Meyer-Lindenberg, "Functional Imaging Research in Schizophrenia," in *International Review of Neurobiology: Neuroimaging, Part B,* Volume 67 ed, M. F. Glabus, Ed.: Academic Press, 2005, pp. 95-118.

[5] M. Breakspear, J. R. Terry, K. J. Friston, A. W. Harris, L. M. Williams, K. Brown, J. Brennan, and E. Gordon, "A disturbance of nonlinear interdependence in scalp EEG of subjects with first episode schizophrenia," *NeuroImage,* vol. 20, pp. 466-478, 09 2003.

[6] K. J. Friston, "The disconnection hypothesis," *Schizophr.Res.,* vol. 30, pp. 115-125, 03/10/ 1998.

[7] G. D. Pearlson and L. Marsh, "Structural brain imaging in schizophrenia: a selective review," *Biological Psychiatry,* vol. 46, pp. 627-649, 09/01/ 1999.

[8] R. M. Murray and S. W. Lewis, "Is schizophrenia a neurodevelopmental disorder?," *Br. Med. J (Clin. Res. Ed),* vol. 295, pp. 681-682, 09/19/ 1987.

[9] D. R. Weinberger, "Implications of normal brain development for the pathogenesis of schizophrenia," *Arch. Gen. Psychiatry,* vol. 44, pp. 660-669, 07 1987.

[10] J. O. Davis and H. S. Bracha, "Prenatal growth markers in schizophrenia: a monozygotic cotwin control study," *Am. J. Psychiatry,* vol. 153, pp. 1166-1172, 09 1996.

[11] A. Lane, A. Kinsella, P. Murphy, M. Byrne, J. Keenan, K. Colgan, B. Cassidy, N. Sheppard, R. Horgan, J. L. Waddington, C. Larkin, and E. O'Callaghan, "The anthropometric assessment of dysmorphic features in schizophrenia as an index of its developmental origins," *Psychol. Med.,* vol. 27, pp. 1155-1164, 09 1997.

[12] P. Jones, B. Rodgers, R. Murray, and M. Marmot, "Child development risk factors for adult schizophrenia in the British 1946 birth cohort," *Lancet,* vol. 344, pp. 1398-1402, 11/19/ 1994.

[13] J. J. Kulynych, L. F. Luevano, D. W. Jones, and D. R. Weinberger, "Cortical abnormality in schizophrenia: an in vivo application of the gyrification index," *Biological Psychiatry,* vol. 41, pp. 995-999, 05/15/ 1997.

[14] L. D. Selemon, G. Rajkowska, and P. S. Goldman-Rakic, "Abnormally high neuronal density in the schizophrenic cortex. A morphometric analysis of prefrontal area 9 and occipital area 17," *Arch. Gen. Psychiatry,* vol. 52, pp. 805-818, 10 1995.

[15] D. W. Zaidel, M. M. Esiri, and P. J. Harrison, "The hippocampus in schizophrenia: lateralized increase in neuronal density and altered cytoarchitectural asymmetry," *Psychol. Med.,* vol. 27, pp. 703-713, 05 1997.

[16] K. J. Friston, "Schizophrenia and the disconnection hypothesis," *Acta Psychiatr. Scand. Suppl,* vol. 395, pp. 68-79, 1999.

[17] Y. Fan, D. Shen, and C. Davatzikos, "Classification of structural images via high-dimensional image warping, robust feature extraction, and SVM," *Med.Image Comput.Comput.Assist.Interv.Int.Conf.Med.Image Comput.Comput.Assist.Interv.,* vol. 8, pp. 1-8, 2005.

[18] K. Nakamura, Y. Kawasaki, M. Suzuki, H. Hagino, K. Kurokawa, T. Takahashi, L. Niu, M. Matsui, H. Seto, and M. Kurachi, "Multiple structural brain measures obtained by three-dimensional magnetic resonance imaging to distinguish between schizophrenia patients and normal subjects," *Schizophr.Bull.,* vol. 30, pp. 393-404, 2004.

[19] V. D. Calhoun, K. A. Kiehl, P. F. Liddle, and G. D. Pearlson, "Aberrant Localization of Synchronous Hemodynamic Activity in Auditory Cortex Reliably Characterizes Schizophrenia," *Biological Psychiatry,* vol. 55, pp. 842-849, 2004.

[20] M. T. Tsuang, N. Nossova, T. Yager, M. M. Tsuang, S. C. Guo, K. G. Shyu, S. J. Glatt, and C. C. Liew, "Assessing the validity of blood-based gene expression profiles for the classification of schizophrenia and bipolar disorder: a preliminary report," *Am. J. Med. Genet. B Neuropsychiatr. Genet.,* vol. 133, pp. 1-5, 02/05/ 2005.

[21] P. J. Pardo, A. P. Georgopoulos, J. T. Kenny, T. A. Stuve, R. L. Findling, and S. C. Schulz, "Classification of adolescent psychotic disorders using linear discriminant analysis," *Schizophr. Res.,* vol. 87, pp. 297-306, 10 2006.

[22] K. J. Friston and C. D. Frith, "Schizophrenia: a disconnection syndrome?," *Clin. Neurosci.,* vol. 3, pp. 89-97, 1995.

[23] M. Liang, Y. Zhou, T. Jiang, Z. Liu, L. Tian, H. Liu, and Y. Hao, "Widespread functional disconnectivity in schizophrenia with resting-state functional magnetic resonance imaging," *Neuroreport,* vol. 17, pp. 209-213, 02/06/ 2006.

[24] K. J. Friston, C. D. Frith, R. S. Frackowiak, and R. Turner, "Characterizing dynamic brain responses with fMRI: a multivariate approach," *NeuroImage,* vol. 2, pp. 166-172, 06 1995.

[25] K. Friston, A. Holmes, K. J. Worsley, J. P. Poline, C. D. Frith, and R. S. Frackowiak, "Statistical parametric maps in functional imaging: A general linear approach," *Hum. Brain Map.,* vol. 2, pp. 189-210, 1995.

[26] G. M. Josin and P. F. Liddle, "Neural network analysis of the pattern of functional connectivity between cerebral areas in schizophrenia," *Biol.Cybern.,* vol. 84, pp. 117-122, 02 2001.

[27] P. F. Liddle, K. J. Friston, C. D. Frith, S. R. Hirsch, T. Jones, and R. S. Frackowiak, "Patterns of cerebral blood flow in schizophrenia," *Br. J. Psychiatry,* vol. 160, pp. 179-186, 02 1992.

[28] C. D. Frith, K. J. Friston, S. Herold, D. Silbersweig, P. Fletcher, C. Cahill, R. J. Dolan, R. S. Frackowiak, and P. F. Liddle, "Regional brain activity in chronic schizophrenic patients during the performance of a verbal fluency task," *Br. J. Psychiatry,* vol. 167, pp. 343-349, 09 1995.

[29] A. Hyvarinen and E. Oja, "Independent component analysis: algorithms and applications," *Neural Netw.,* vol. 13, pp. 411-430, 05 2000.

[30] J. Bell and T. J. Sejnowski, "An information maximisation approach to blind separation and blind deconvolution," *Neural Comput.,* vol. 7, pp. 1129-1159, 1995.

[31] M. J. McKeown, S. Makeig, G. G. Brown, T. P. Jung, S. S. Kindermann, A. J. Bell, and T. J. Sejnowski, "Analysis of fMRI Data by Blind Separation Into Independent Spatial Components," *Hum. Brain Map.,* vol. 6, pp. 160-188, 1998.

[32] V. D. Calhoun and T. Adali, "'Unmixing' Functional Magnetic Resonance Imaging with Independent Component Analysis," *IEEE Eng. in Medicine and Biology,* vol. 25, pp. 79-90, 2006.

[33] J. R. Duann, T. P. Jung, W. J. Kuo, T. C. Yeh, S. Makeig, J. C. Hsieh, and T. J. Sejnowski, "Single-Trial Variability in Event-Related BOLD Signals," *Neuroimage.,* vol. 15, pp. 823835, 04 2002.

[34] V. D. Calhoun, T. Adali, G. D. Pearlson, and J. J. Pekar, "A Method for Making Group Inferences from Functional MRI Data Using Independent Component Analysis," *Hum. Brain Map.,* vol. 14, pp. 140-151, 03/01/ 2001.

[35] F. Beckmann, M. De Luca, J. T. Devlin, and S. M. Smith, "Investigations into resting-state connectivity using Independent Component Analysis," *Philos.Trans. R. Soc. Lond. B. Biol. Sci.,* vol. 360, pp. 1001-1013, 2005.

[36] A.Garrity, G. D. Pearlson, K. McKiernan, D. Lloyd, K. A. Kiehl, and V. D. Calhoun, "Aberrant 'default mode' functional connectivity in schizophrenia," *Am. J.Psychiatry,* 2006.

[37] M. D. Greicius, G. Srivastava, A. L. Reiss, and V. Menon, "Default-mode network activity distinguishes Alzheimer's disease from healthy aging: evidence from functional MRI," *Proc. Natl. Acad. Sci. U.S.A,* vol. 101, pp. 4637-4642, 03/30/ 2004.

[38] M. D. Greicius and V. Menon, "Default-mode activity during a passive sensory task: uncoupled from deactivation but impacting activation," *J. Cogn. Neurosci,* vol. 16, pp. 14841492, Nov 2004.

[39] J. C. Rajapakse, C. L. Tan, X. Zheng, S. Mukhopadhyay, and K. Yang, "Exploratory analysis of brain connectivity with ICA," *IEEE Eng. Med. Biol. Mag.,* vol. 25, pp. 102-111, 03 2006.

[40] R. E. Passingham, S. E. Klaas, and R. Kotter, "The Anatomical Basis of Functional Localization in the Cortex," *Nat. Neurosci.,* vol. 3, pp. 606-616, 2002.

[41] Horwitz, "The elusive concept of brain connectivity," *NeuroImage,* vol. 19, pp. 466-470, 06 2003.

[42] N. Ramnani, L. Lee, A. Mechelli, C. Phillips, A. Roebroeck, and E. Formisano, "Exploring brain connectivity: a new frontier in systems neuroscience. Functional Brain Connectivity, 4-6 April 2002, Dusseldorf, Germany," *Trends Neurosci.,* vol. 25, pp. 496-497, 10 2002.

[43] K. J. Friston, C. D. Frith, P. F. Liddle, and R. S. Frackowiak, "Functional connectivity: the principal-component analysis of large (PET) data sets," *J. Cereb. Blood Flow Metab,* vol. 13, pp. 5-14, 01 1993.

[44] Buchel, J. T. Coull, and K. J. Friston, "The predictive value of changes in effective connectivity for human learning," *Science,* vol. 283, pp. 1538-1541, 03/05/ 1999.

[45] R. Schlosser, T. Gesierich, B. Kaufmann, G. Vucurevic, S. Hunsche, J. Gawehn, and P. Stoeter, "Altered effective connectivity during working memory performance in schizophrenia: a study with fMRI and structural equation modeling," *Neuroimage.,* vol. 19, pp. 751-763, 07 2003.

[46] V. D. Calhoun, G. D. Pearlson, and K. A. Kiehl, "Neuronal Chronometry of Target Detection: Fusion of Hemodynamic and Event-related Potential Data," *NeuroImage,* vol. 30, pp. 544-553, 2006.

[47] J. S. George, C. J. Aine, J. C. Mosher, D. M. Schmidt, D. M. Ranken, H. A. Schlitt, C. C. Wood, J. D. Lewine, J. A. Sanders, and J. W. Belliveau, "Mapping function in the human brain with magnetoencephalography, anatomical magnetic resonance imaging, and functional magnetic resonance imaging," *J. Clin. Neurophysiol.,* vol. 12, pp. 406-431, 09 1995.

[48] R. G. Schlosser, I. Nenadic, G. Wagner, D. Gullmar, K. von Consbruch, S. Kohler, C. C. Schultz, K. Koch, C. Fitzek, P. M. Matthews, J. R. Reichenbach, and H. Sauer, "White matter abnormalities and brain activation in schizophrenia: A combined DTI and fMRI study," *Schizophr. Res,* vol. 89, pp. 1-11, Jan 2007.

[49] V. D. Calhoun, T. Adali, N. Giuliani, J. J. Pekar, G. D. Pearlson, and K. A. Kiehl, "A Method for Multimodal Analysis of Independent Source Differences in Schizophrenia: Combining Gray Matter Structural and Auditory Oddball Functional Data," *Hum. Brain Map.,* vol. 27, pp. 47-62, 2006.

[50] M. M. Mesulam, "From sensation to cognition," *Brain,* vol. 121 ( Pt 6), pp. 1013-1052, 06 1998.

[51] Bullmore, M. Brammer, S. C. Williams, V. Curtis, P. McGuire, R. Morris, R. Murray, and T. Sharma, "Functional MR imaging of confounded hypofrontality," *Hum. Brain Map.,* vol. 8, pp. 86-91, 1999.

[52] K. R. Laurens, E. T. Ngan, A. T. Bates, K. A. Kiehl, and P. F. Liddle, "Rostral anterior cingulate cortex dysfunction during error processing in schizophrenia," *Brain,* vol. 126, pp. 610-622, 03 2003.

[53] D. S. Manoach, R. L. Gollub, E. S. Benson, M. M. Searl, D. C. Goff, E. Halpern, C. B. Saper, and S. L. Rauch, "Schizophrenic subjects show aberrant fMRI activation of

dorsolateral prefrontal cortex and basal ganglia during working memory performance," *Biological Psychiatry,* vol. 48, pp. 99-109, 07/15/ 2000.

[54] J. Schroder, M. Essig, K. Baudendistel, T. Jahn, I. Gerdsen, A. Stockert, L. R. Schad, and M. V. Knopp, "Motor dysfunction and sensorimotor cortex activation changes in schizophrenia: A study with functional magnetic resonance imaging," *NeuroImage,* vol. 9, pp. 81-87, 01 1999.

[55] A R. McIntosh and F. Gonzalez-Lima, "Structural Equation Modeling and Its Application to Network Analysis in Functional Brain Imaging," *Hum. Brain Map.,* vol. 2, pp. 2-22, 1994.

[56] K. J. Friston, L. Harrison, and W. Penny, "Dynamic causal modelling," *NeuroImage,* vol. 19, pp. 1273-1302, 08 2003.

[57] M. N. Rajah and A. R. McIntosh, "Overlap in the functional neural systems involved in semantic and episodic memory retrieval," *J. Cogn. Neurosci.,* vol. 17, pp. 470-482, 03 2005.

[58] A. Meyer-Lindenberg, P. Kohn, C. B. Mervis, J. S. Kippenhan, R. K. Olsen, C. A. Morris, and K. F. Berman, "Neural basis of genetically determined visuospatial construction deficit in williams syndrome," *Neuron,* vol. 43, pp. 623-631, 09/02/ 2004.

[59] R. McIntosh, F. L. Bookstein, J. V. Haxby, and C. L. Grady, "Spatial pattern analysis of functional brain images using partial least squares," *NeuroImage,* vol. 3, pp. 143-157, 06 1996.

[60] K. Friston, J. P. Poline, S. Strother, A. Holmes, C. D. Frith, and R. S. Frackowiak, "A Multivariate analysis of PET activation studies," *Hum. Brain Map.,* vol. 4, pp. 140-151, 1996.

[61] V. D. Calhoun, T. Adali, and J. Liu, "A Feature-based Approach to Combine Functional MRI, Structural MRI, and EEG Brain Imaging Data," in *Proc. EMBS,* 2006.

[62] K. A. Kiehl, M. C. Stevens, K. Celone, M. Kurtz, and J. H. Krystal, "Abnormal hemodynamics in schizophrenia during an auditory oddball task," *Biological Psychiatry,* vol. 57, pp. 1029-1040, 05/01/ 2005.

[63] M. R. Johnson, N. Morris, R. S. Astur, V. D. Calhoun, D. H. Mathalon, K. A. Kiehl, and G. D. Pearlson, "Investigation of Working Memory Abnormalities in Schizophrenia: An fMRI Study," *Biological Psychiatry,* 2004.

[64] N. C. Andreasen, S. Paradiso, and D. S. O'Leary, ""Cognitive dysmetria" as an integrative theory of schizophrenia: a dysfunction in cortical-subcortical-cerebellar circuitry?," *Schizophr. Bull.,* vol. 24, pp. 203-218, 1998.

[65] P. F. Liddle, "PET scanning and schizophrenia--what progress?," *Psychol. Med.,* vol. 22, pp. 557-560, 08 1992.

[66] L. Friedman and G. H. Glover, "Report on a multicenter fMRI quality assurance protocol," *J. Magn Reson. Imaging,* vol. 23, pp. 827-839, 06 2006.

[67] R. K. Heinssen, B. N. Cuthbert, J. Breiling, L. J. Colpe, and R. Dolan-Sewell, "Overcoming barriers to research in early serious mental illness: issues for future collaboration," *Schizophr. Bull,* vol. 29, pp. 737-745, 2003.

[68] J. A. Lieberman, G. D. Tollefson, C. Charles, R. Zipursky, T. Sharma, R. S. Kahn, R. S. Keefe, A. I. Green, R. E. Gur, J. McEvoy, D. Perkins, R. M. Hamer, H. Gu, and M. Tohen, "Antipsychotic drug effects on brain morphology in first-episode psychosis," *Arch. Gen. Psychiatry,* vol. 62, pp. 361-370, Apr 2005.

[69] D. L. Braff, R. Freedman, N. J. Schork, and Gottesman, II, "Deconstructing schizophrenia: an overview of the use of endophenotypes in order to understand a complex disorder," *Schizophr. Bull,* vol. 33, pp. 21-32, Jan 2007.

[70] A. R. Hariri, E. M. Drabant, and D. R. Weinberger, "Imaging genetics: perspectives from studies of genetically driven variation in serotonin function and corticolimbic affective processing," *Biological Psychiatry,* vol. 59, pp. 888-897, 05/15/ 2006.

[71] P. J. Harrison and D. R. Weinberger, "Schizophrenia genes, gene expression, and neuropathology: on the matter of their convergence," *Mol. Psychiatry,* vol. 10, pp. 40-68, 2004/07/20/online 2004.

[72] T. E. Goldberg and D. R. Weinberger, "Genes and the parsing of cognitive processes," *Trends Cogn. Sci.,* vol. 8, pp. 325-335, 07 2004.

[73] McDonald, N. Marshall, P. C. Sham, E. T. Bullmore, K. Schulze, B. Chapple, E. Bramon, F. Filbey, S. Quraishi, M. Walshe, and R. M. Murray, "Regional brain morphometry in patients with schizophrenia or bipolar disorder and their unaffected relatives.," *Am. J. Psychiatry,* vol. 163, pp. 478-487, 2006.

[74] Gottesman, II and D. R. Hanson, "Human development: biological and genetic processes," *Annu. Rev. Psychol,* vol. 56, pp. 263-286, 2005.

[75] M. F. Egan, T. E. Goldberg, B. S. Kolachana, J. H. Callicott, C. M. Mazzanti, R. E. Straub, D. Goldman, and D. R. Weinberger, "Effect of COMT Val108/158 Met genotype on frontal lobe function and risk for schizophrenia," *Proceedings of the National Academy of Sciences of the United States of America,* vol. 98, pp. 6917-6922, 2001.

[76] T. E. Goldberg, M. F. Egan, T. Gscheidle, R. Coppola, T. Weickert, B. S. Kolachana, D. Goldman, and D. R. Weinberger, "Executive subprocesses in working memory: relationship to catechol-O-methyltransferase Val158Met genotype and schizophrenia," *Archives of General Psychiatry,* vol. 60, pp. 889-896, Sep 2003.

[77] J. H. Callicott, R. E. Straub, L. Pezawas, M. F. Egan, V. S. Mattay, A. R. Hariri, B. A. Verchinski, A. Meyer-Lindenberg, R. Balkissoon, B. Kolachana, T. E. Goldberg, and D. R. Weinberger, "Variation in DISC1 affects hippocampal structure and function and increases risk for schizophrenia," *PNAS,* vol. 102, pp. 8627-8632, June 14, 2005 2005.

*Chapter 7*

# TRIGEMINO-CARDIAC REFLEX AND IMAGING – THE PROMISE OF NEW INSIGHTS

## *B. Schaller, A. Filis and M. Buchfelder*
Departments of Neurosurgery,
University Hospitals Erlangen-Nurnberg and Munster,
Germany

The trigemino-cardiac reflex (TCR) is a well-known entity that has gained more and more interest [1]. The reproducible hypotension and bradycardia upon stimulation of the trigeminal nerve, has been reported during craniofacial surgery and during surgery within the cerebellopontine angle, petrosal sinus, orbit, trigeminal ganglion and the falx. The stimulation of a sensory branch of the trigeminal nerve results in the hyperactivity of the trigeminal ganglion, thereby triggering the TCR [1]. The dorsal region of the spinal trigeminal tract includes neurons from hypoglossal and vagus nerves, and projections have been seen between the vagus and trigeminal nuclei. The vagus provides parasympathetic innervation to the heart, vascular smooth muscle, and abdominal viscera. Vagal stimulation via these connections after trigeminal nerve activation likely accounts for the reflexive response of asystole seen in this patient. This is confirmed by the observation that the reflex was inhibited by the anticholinergic effects of glycopyrrolate. Awareness of TCR allows for early detection and appropriate treatment.

As soon as the TCR occurs, an interruption of the surgical manoeuvres is sufficient for returning to normal hemodynamic parameters. Major neurological complications related to intra-operative hypotension due to TCR were observed in different studies in the cerebellopontine angle.

Radiological surrogate markers of potential anoxic/hypoxic cerebral damage may be of importance in a second step [2]. Each serum or radiological marker has its pros and cons. To accurately prognosticate the occurrence of TCR, a multimodal scale or algorithm that incorporates serum biomarkers as well as (intraoperative) radiological markers may be needed. As these techniques are being evaluated more closely and as imaging modalities increase in sensitivity and portability, we will continue to provide some guidance as to which patients have no chance of meaningful recovery after TCR. This means that we may have certain preoperative constellations of radiological and serum surrogate markers in which there is a great chance of intraoperative occurrence of TCR. But until iintraoperative imaging modalities can routinely be used to image the brain at a cellular level, MRI will always be secondary to pathophyisological events, such as the occurrence of the TCR, in the final diagnostic evaluation.

Efforts to unravel the mechanisms of this instability following (transient) ischemic events, such as seen in TCR, using imaging studies have led to new concepts and definitions, and sparked further debate. While imaging has increased diagnostic certainty, it has yet to provide reliable prognostic markers. The evidence suggests that risk of clinical recurrence is most closely linked to the degree to which the initial deficit reverses. From a tissue level, however, there are also data to support the notion of a 'stroke-prone state' following both transient ischemic attack and completed stroke, suggesting that mechanistically they may be less distinct than previously thought. Transient ischemic events, such as the TCR, may simply highlight the dynamic nature of all acute ischemic cerebrovascular syndromes.

A general consensus is that combinations of imaging methods will ultimately be most fruitful in predicting disease. Their roll-out into translational practice will not be free of complexity, however, as values differ in terms of what defines benefit and risk, who will benefit and who is at risk, what methods must be in place to assure the maximum safety, comfort, and protection of subjects and patients, and educational and policy needs.

## REFERENCES

[1] Schaller B, Probst R, Strebel S, Gratzl O. Trigeminocardiac reflex during surgery in the cerebellopontine angle. *J. Neurosurg.* 1999 Feb;90(2):215-20

[2] Schaller B, Buchfelder M. The Janus Shape of Glucose and Brain Under Stress Conditions. *Neuroscience Imaging* 2006; 1: 229-230.

Chapter 8

# THE ROLE OF IMAGING IN NEUROLOGICAL FACIAL PAIN

### S. E. J. Connor[*]
Department of Neuroradiology,
King's College Hospital,
Denmark Hill, London SE5 9RS, UK;

## ABSTRACT

This review examines the application of imaging to the neurological aspects of facial pain. Neuralgic facial pain focusing on trigeminal neuralgia, facial pain with associated cranial nerve deficits, pain referred through shared roots of cranial nerve innervation, and other neurological facial pain syndromes will be discussed. The diagnostic accuracy, diagnostic yield of imaging and the impact on clinical management in these clinical settings will be reviewed and the structural correlates of facial pain will be illustrated.

## INTRODUCTION

There are numerous and varied causes of pain in the facial area and it represents a considerable diagnostic challenge to the physician. Specific aspects of the clinical presentation may be used to categorise facial pain and guide appropriate imaging protocols. This review will focus on neurological aspects of facial pain and it is useful to divide the discussion into neuralgias, facial pain with cranial nerve deficits, pure facial pain referred by shared cranial nerve innervation and the role of imaging in other neurological facial pain syndromes. It should be considered that these entities are much less common than non neurological causes of pure facial pain such as dental disease, paranasal sinus disease, and temporo-mandibular joint disease which are beyond the scope of this review [1,2]. For each category, appropriate imaging protocols and techniques will be discussed together with

---

[*] Email: sejconnor@tiscali.co.uk

relevant anatomical aspects, possible pathologies and available data on the diagnostic accuracy and yield of imaging in this clinical setting.

## NEURALGIA

Neuralgia is characterised by sudden intense sharp lancating burning or stabbing pain, which lasts for less than 2 minutes and is recurring. It is often triggered by a sensory or mechanical stimulae and there should really be no neurological deficit in classic trigeminal neuralgia [3]. Trigeminal neuralgia is the most frequent neuralgia, however the prevalence remains only 4-5 per hundred thousand [4]. Trigeminal neuralgia is characterised by pain located in the distribution of the second and third, and occasionally first, divisions of the trigeminal nerves [5]. John Fothergill gave the first full and accurate description of trigeminal neuralgia in 1773 and it may be referred to as Fothergills disease [6].

In the absence of an underlying structural lesion, there is still no evidence for the pathophysiological mechanisms of trigeminal neuralgia. The most widespread theory is of vascular compression of the root entry zone of the trigeminal nerve, adjacent to the pons, where there is less organised myelin and parallel nerve fibres. It is speculated that this promotes ectopic stimulation from the altered nerve fibres and allows transmission of painful impulses [7, 8]. Since there are no satisfactory animal models, the hypothesis is based on the idiosyncratic nature of the clinical findings and response to various therapies. More recently, advances in the understanding of the electrical behaviour in injured sensory neurons, together with histopathological observations, has led to the ignition hypothesis [9]. This suggests that the symptoms arise from specific abnormality in the trigeminal root or ganglion with hyper-excitable axons, which results in synchronised after discharge activity. This theory implies that demyelination itself may not be responsible and that the abnormality may be anywhere along the trigeminal nerve root or ganglion.

Patients who cannot tolerate the adverse affects of, or are refractory to, the pharmaceutical treatments, are usually recommended for surgery. The only surgical treatment modality that directly addresses the presumed underlying pathology is microvascular decompression (MVD), which has been promoted by Jannetta since 1967 [7]. Contemporary microsurgery has allowed surgeons to operate with greater confidence and safety within the posterior fossa. This has been the most successful treatment with the lowest rate of symptom recurrence. MVD generally preserves sensation and achieves postoperative immediate pain relief in 91 to 97% of cases, with long term efficacy in between 57 to 70% of cases [10, 11]. There are a large number of reports, which assess the accuracy of magnetic resonance imaging (MRI) in predicting the compression of the trigeminal nerve root pre-operatively. Neurosurgeons will gain information about the presence of vessels and this will help to determine how to explore these areas.

Neurovascular compression has been evaluated by using thin section source images and reformatted images of MR angiographic sequences. Approaches by different MRI manufacturers include 3D fast imaging with study state precession (or FISP) or 3D spoiled gradient recalled acquisition in the study state (or GRASS) (Figure 1). An artery with fast blood flow is seen as high signal intensity, whilst the nerve is of intermediate signal intensity. However contrast resolution between CSF and the nerve is unclear and the depiction of veins is poor due to slow flow unless gadolinium is administered. High spatial resolution 3D imaging with constructive interference (CISS) has also been investigated in order to provide high resolution, heavily T2 weighted images with flow compensation (Figure 2).

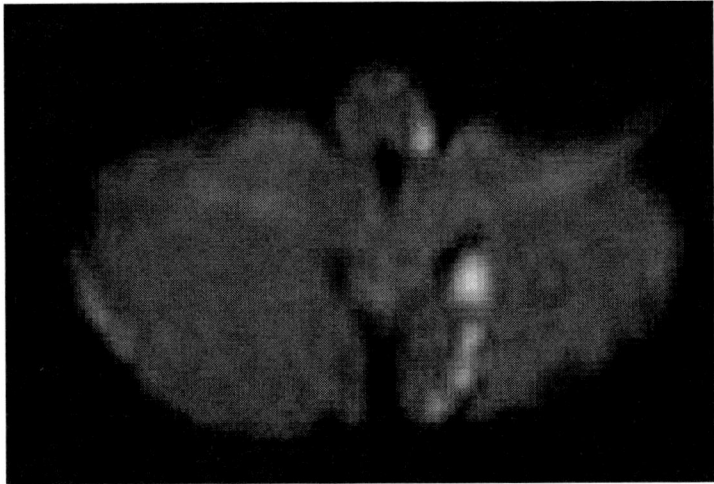

Figure 1. Patient with right trigeminal neuralgia. GRASS axial image demonstrates vascular compression of the right trigeminal nerve root entry zone (white arrowhead) with distortion and atrophy of the nerve.

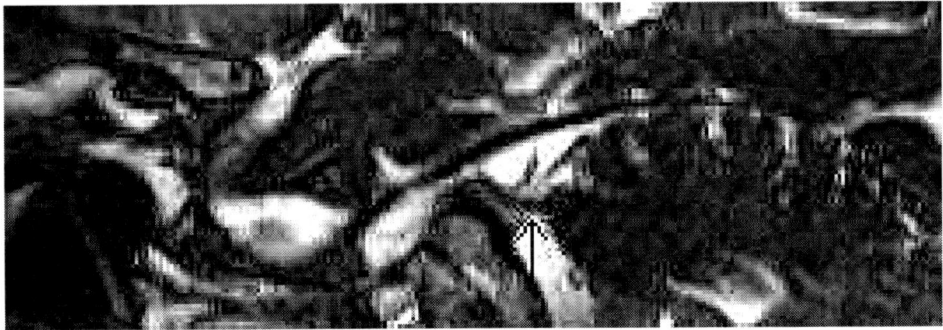

Figure 2. Patient with right trigeminal neuralgia. Reformatted CISS sagittal image demonstrates grooving of the right trigeminal root entry zone by a vessel (black arrow).

This is capable of contrasting the vascular structures, both veins and arteries from the nerve within the CSF.

Evaluating the diagnostic accuracy of these sequences from the available literature is problematic [12-23]. The diagnostic criteria for trigeminal neuralgia are usually not stated and there are often single observers who are unblended. Not all patients have surgical validation and there is little information on the significance of imaging findings in terms of the success of surgical treatment [24-26]. Some sensitivities are stated relative the patients symptoms and some relative to surgical findings, whilst some studies use the contralateral to the symptoms as a control, whereas others use non-trigeminal nerve control patients to determine specificity. Sensitivity and specificity are wide ranging between 50 and a 100%. Clearly the assessment of sensitivity related to symptoms may be flawed if the ignition hypothesis [9] is true, since there may be pathology related to the trigeminal ganglion rather than the trigeminal nerve roots.

Assessing for neurovascular compression is only important if the recommendation for treatment of a patient with trigeminal neuralgia depends on the demonstration of a compressing vessel pre-operatively. This would be the case if more minimally invasive peripheral procedures, such as radio frequency thermo-coagulation or balloon compression of the peripheral nerve or ganglion were to be used in preference if MRI did not reveal neurovascular contact [27]. The sensitivity of MRI techniques appears less important in some surgical practices. Some surgeons will still perform a craniotomy even if there is no evidence of neurovascular contact on MRI and the surgeon will perform a partial sensory rhizotomy if no vessel is found compressing the nerve at surgery. Some surgeons have described operative findings when no neurovascular contact was present at operation. Features such as membranes impinging on the nerve, or short cisternal segments, have been described [28]. The detection of neurovascular contact on MRI may also be of benefit in predicting response to gamma knife treatment [24-26].

It should be noted that neurovascular contact is seen in a significant proportion of asymptomatic patients and sides. Autopsy series of previously asymptomatic individuals has revealed what was deemed to be significant compression in between 32 and 58% of cases [29-31]. Therefore certain other anatomical features have been studied to refine the judgement as to whether the contact is significant. Grooving, distortion and nerve atrophy (Figure 1) has led to increased specificity and other authorities have referred to vessel running perpendicular to the nerve as being features associated with symptomatic neurovascular contact [18]. Contact is felt to be most significant if it occurs within the most central portion of the nerve or so called root entry zone. In this portion there is less organised myelin and parallel running nerve fibres, which are prone to injury. The length of nerve covered by such central myelin is variably stated and in many studies is not stated at all. A figure of 6mm from the pons is stated to be the upper limit [32, 33]. Arterial contact is the usual source of neurovascular compression, with the superior cerebellar artery being most frequently implicated. Occasionally there is more striking distortion of the nerves at the root entry zone in the setting of trigeminal neuralgia due to compression by vertebro-basilar ectasia (Figure 3), or possibly a foetal type posterior communicating artery. Veins may however contribute to neurovascular compression and they contributed to approximately an eighth of cases in the Jannetta series [11]. Venous decompression appears to be almost as effective as that for arteries, although there is an increased risk of symptomatic recurrence. It should be considered that a static image is being viewed and that displacement of the nerve may vary with cardiac cycle. This issue has not yet been explored with cardiac gating.

Since there are frequent neurovascular contacts in control subjects and sides it appears that contact is frequently necessary but not sufficient condition for the development of trigeminal neuralgia. Since there are no definite criteria separating significant from incidental compression, the visualisation of neurovascular contact should not be used as a convenient way of establishing a diagnosis, as satisfactory outcomes are still best guided by the clinical diagnosis based on sound diagnostic criteria. The risk of using MRI findings of neurovascular contact for diagnosis applies in particular to other chronic orofacial pains, which are estimated to occur by some in up to 23% of the population [34]. Since such pain syndromes are ten thousand times more common than trigeminal neuralgia [4, 34], an incidental association of neurovascular contact with orofacial pain will be considerably more frequent than a pathophysiological relationship of neurovascular contact with trigeminal neuralgia.

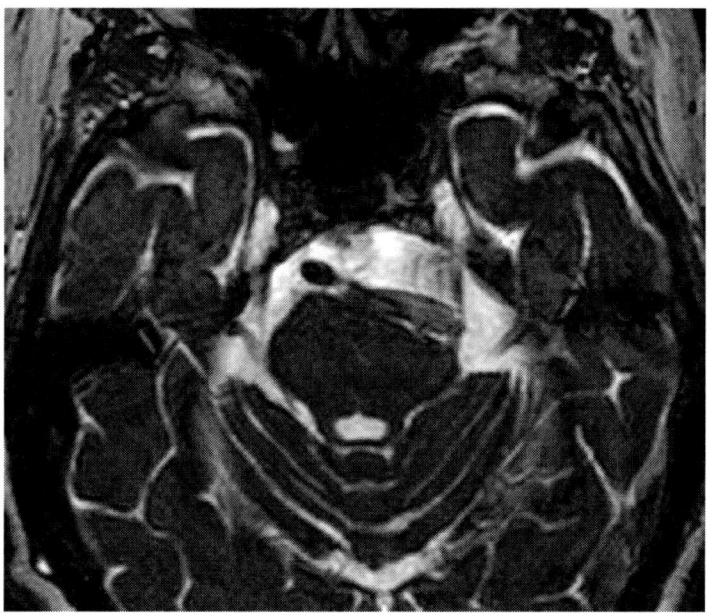

Figure 3. Patient with left sided facial pain. Axial CISS image reveals neurovascular contact at the left trigeminal nerve root entry zone by an ectatic basilar artery.

Trigeminal neuralgia may occasionally be secondary to another structural abnormality. Since classic primary trigeminal neuralgia is less common in younger patients, is rarely bilateral, and classic descriptions define no sensory impairment, these findings should raise the possibility of secondary trigeminal neuralgia. Secondary trigeminal neuralgia may be due to benign or malignant tumours of the posterior fossa. These are felt to result in trigeminal neuralgia by direct tumoural compression, arterial displacement with neurovascular compression or chemical irritation by neoplastic factors [35]. It has been observed that contact rather than infiltration is more likely to cause trigeminal neuralgia [36]. Multiple sclerosis should be considered as a cause of secondary trigeminal neuralgia, typically due to a plaques arising adjacent to the trigeminal nerve root entry zone.

Historically, it has been accepted that 15% of patients with trigeminal neuralgia have secondary or symptomatic trigeminal neuralgia [37]. The largest series which studied the incidence of tumours as a secondary cause, is that of Cheng et al who included a large of number of patients with trigeminal neuralgia with follow up for up to 15 years [38]. Ten percent of these were subsequently found to have tumours, however it should be noted that only a minority of these patients had classic trigeminal neuralgia throughout the follow up period.

The tumours responsible for trigeminal neuralgia are most frequently meningiomas and acoustic neuromas [38, 39] (Figure 4) although it remains an infrequent clinical presentation. A significantly higher incidence of trigeminal neuralgia is associated with cerebellopontine cistern epidermoids and some small surgical series have shown 70 to 100% of such patients to develop trigeminal neuralgia [35, 40-41]. Occasionally lesions sited more peripherally along the trigeminal nerve pathway such as metastases (Figure 5) or perineural malignancy may result in neuralgic type facial pain without detectable sensory loss [42].

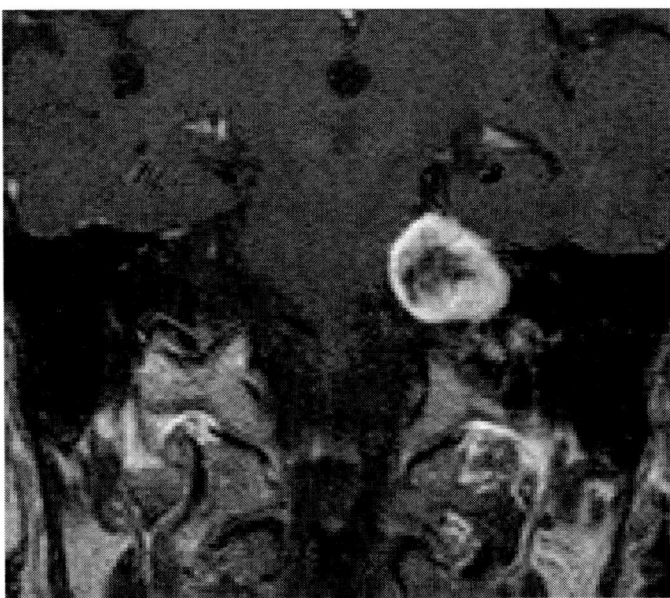

Figure 4. Patient with known acoustic neuroma developed facial pain on follow up. Post gadolinium coronal T1-w image demonstrates an acoustic neuroma extending superiorly to the left trigeminal nerve root entry zone.

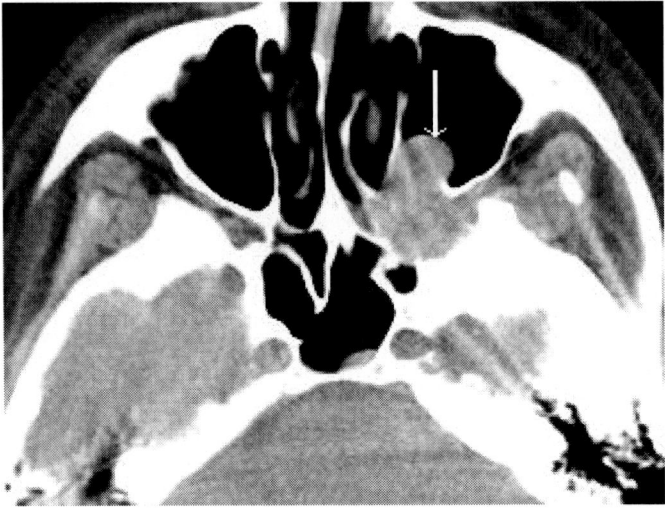

Figure 5. Patient with left sided trigeminal neuralgia without sensory loss. Axial CT scan shows a soft tissue mass centred on the left pterygopalatine fossa and eroding the posterior maxillary antrum (white arrow). This resulted from a metastasis from an endometrial carcinoma.

No good data on the incidence of multiple sclerosis in patients presenting with trigeminal neuralgia is available, however the presence of multiple sclerosis is a risk factor for trigeminal neuralgia which is frequently refractory to treatment [43]. One study identified almost 2% of this population of patients to experience trigeminal neuralgia, however in only a small proportion of these was it present at the initial diagnosis [44]. The patients with brain stem lesions in the region of the root entry zone, such as infarction [45], angiomas and arteriovenous malformations [46], may also occasionally develop trigeminal neuralgia, but rarely in isolation. Occasionally lesions elsewhere within the trigeminal spinal tract or nucleus may develop neuralgic type pain.

Thus MRI is the imaging modality of choice in patients with trigeminal trigeminal neuralgia. Whilst imaging for neurovascular contact with specific MRI sequences should not be used for diagnosis it may be a guide to appropriate surgical therapy if there is failed medical management. Other structural abnormalities are rare in the setting of classical trigeminal neuralgia, however MRI is generally justified, and particularly if there is atypical clinical features or if surgery is considered.

Glossopharyngeal neuralgia is rarer and characterised by paroxysmal neuralgic pain localised to the throat near the tongue or tonsillar fossa and radiating to the jaw or neck [47]. Underlying structural lesions are very uncommon, however it has been associated with vascular compression of the glossopharyngeal nerve by the PICA, and this has been responsive to microvascular decompression surgery [48]. Other neuralgias such as geniculate, Sluders or supraorbital neuralgia do not have imaging correlates. One condition, which sometimes mimics glosso-pharyngeal neuralgia is that of Eagle syndrome (Figure 6) due to an elongated styloid process or calcified stylohyoid ligament [49].

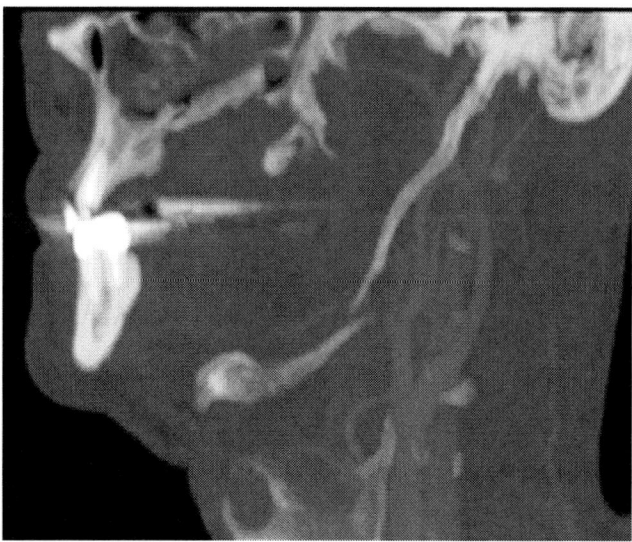

Figure 6. Sagittal reformat of a contrast enhanced CT study imaged on bone windows. There is an elongated calcified stylohyoid ligament in this patient with Eagle syndrome.

## FACIAL PAIN WITH CRANIAL NERVE SIGNS

Facial pain may be associated with eye signs, with disturbed sensation to the face or masticator muscle weakness, with facial palsy or hearing disturbance and with dysphasia, dysphonia or dysarthria.

With regard to eye signs, there are several ocular disorders associated with facial pain, such as acute glaucoma or temporal arteritis, which are diagnosed on clinical criteria and other non imaging investigations. Optic neuritis is a relatively common neurological disorder, which often presents with decreased visual acuity and pain. Optic neuritis is most frequently secondary to multiple sclerosis or other inflammatory or infectious diseases, such as sarcoid or HIV [50]. Imaging of the brain in the setting of optic neuritis may be required and has been shown to predictive of multiple sclerosis in a large population based prospective series. Clinically silent white matter lesions suggestive of multiple sclerosis (MS) are seen in the majority of patients who progressed to MS, and if there are greater than three lesions, then there is a 50% 5 year risk of clinically definite multiple sclerosis [51]. This may have implications in the use of early disease modifying treatments. MRI is often diagnostic, showing increased STIR signal and enhancement or dilatation of the optic nerve sheath in the acute phase [52].

The syndrome of painful ophthalmoplegia incorporates pain in the eye, orbit or forehead and associated with ocular palsies. Various pathologies located in the orbit, orbital apex, cavernous sinus and subarachnoid space may result in this syndrome, and we need to look at these regions on imaging. Inflammatory, neoplastic or vascular causes are possible and CT or MRI is mandatory in this clinical setting.

Non-specific inflammation or idiopathic inflammatory pseudo tumour may involve the uvea, sclera, extraocular muscles, lacrimal gland, optic and ocular motor nerves. When involving the orbital apex or cavernous sinus, this is referred to as Tolosa Hunt syndrome, and there may be associated trigeminal sensory impairment. Orbital infection due to bacteria, viruses, TB or fungus, should also be a major consideration in the setting of painful ophthalmoplegia particularly in immunocompromised patients. Other inflammations, for instance those due to Wegeners granulomatosis, sarcoid, SLE and dermatomyositis are less common.

Painful diplopia or visual disturbance may be secondary to neoplastic disease of the orbit or cavernous sinus. A range of malignancies may be implicated including metastasis, meningioma (Figure 7), sarcoma, perineural malignancy or craniopharyngioma.

Pituitary apoplexy may also present with retro-orbital pain with deteriorating vision and diplopia.

Vascular causes of painful ophthalmaplegia include aneurysm compression, usually due to impingement on the oculomotor nerve by a posterior communicating artery aneurysm. CTA has proved sufficient as a first line of investigation in this setting [53] (Figure 8).

Other vascular lesions, such as carotid dissection, may result in pain referred to the face and associated eye signs, such as Horners syndrome, due to local compression of the sympathetic nerve fibres in the carotid wall. They may also result in visual disturbance due to embolic complications [54].

Facial pain may also be associated with disturbed sensation to the face, or masticator muscle weakness. When there is sensory disturbance, such as numbness or parasthesia to the face up to 70% of patients may demonstrate abnormalities on MRI [55]. There should be a particularly high index of suspicion if there is progression of symptoms or signs of trigeminal neuropathy or a duration of less than a year. The so called numb chin syndrome has an extremely high yield of up to 90%, over half of which have pathology within the mandible [56]. This group of patients may have abnormalities along the trigeminal pathway, including the trigeminothalamocortical tracts, trigeminal nuclei, trigeminal nerve ganglion and the distal extra-cranial divisions [57]. MRI is the imaging modality of choice in this setting and since clinical localisation is poor the entire course of the trigeminal nerve should be visualised.

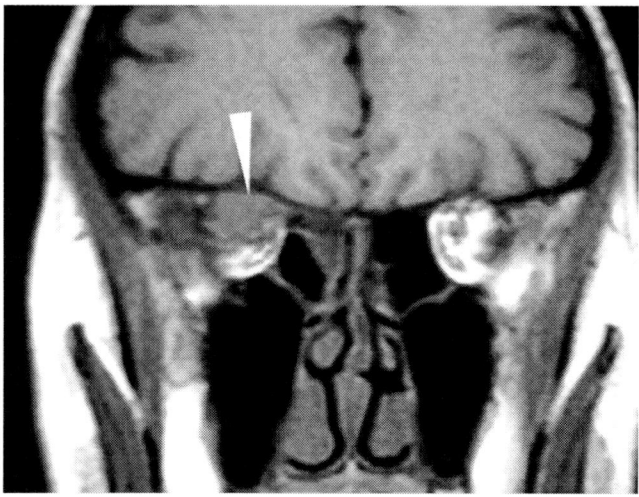

Figure 7. Patient with painful right sided ophthalmoplegia. Coronal T1-w image demonstrates intermediate T1-w signal at the superior right orbital apex (white arrowhead) corresponding to a meningioma.

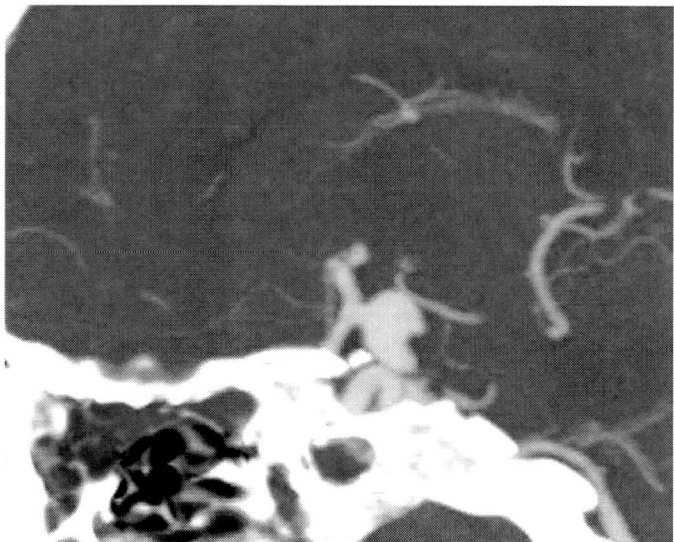

Figure 8. A sagittal reformat of a CTA study demonstrates a bilobed posterior communicating artery aneurysm.

Perineural spread of malignancy is an important cause of facial pain [42] and trigeminal neuropathy. It is typically secondary to adenoid cystic carcinoma, although may be seen in the presence of other tumours, such as squamous cell carcinoma or melanoma. The region of the parapharyngeal fat should always be inspected on MRI or CT, as this forms a crossroads to subsequent spread along adjacent nerves.

Inflammatory lesions such as viral infection, pseudotumour or granulomatous disease may also be implicated and may result in focal masses or perineural spread along the trigeminal nerve and its divisions. In clinical studies where MRI is negative, auto-immune or viral inflammation and micro-angiopathy are felt to be the likely causes of trigeminal neuropathy [58]. Facial pain due to reactivation of trigeminal herpes zoster is a common clinical problem [59]. This is rarely associated with MRI changes within the trigeminal cistern (Figure 9) or trigeminal nucleus [60].

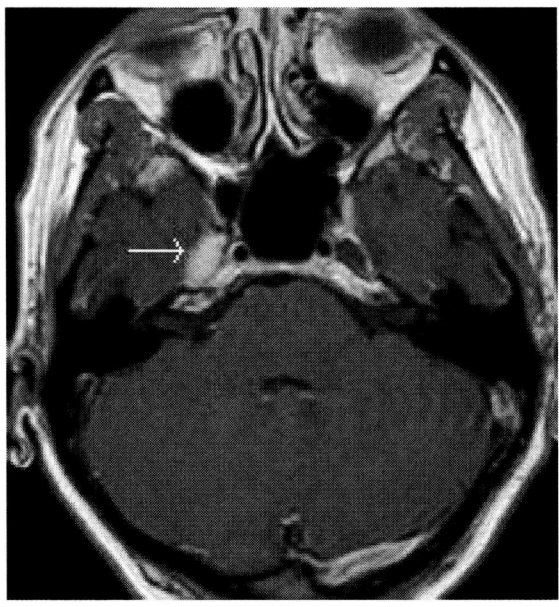

Figure 9. Axial post gadolinium T1-w image reveals enhancement of Meckel's cave (white arrow) in a patient with right sided trigeminal herpes zoster and facial pain.

Facial pain may also occur in conjunction with facial nerve and vestibulocochlear dysfunction. Idiopathic bells palsy or less commonly Ramsey Hunt syndrome are disorders often associated with facial pain around the ear, jaw and neck [61]. There is little role for imaging in such suspected viral neuronitis in the acute stage, although there is some evidence that the degree of contrast enhancement at this stage may predict outcome [62]. The value of imaging generally confined to patients with persistent or progressive facial nerve palsy in order to detect additional structural lesions. This should be performed with MRI post gadolinium with thin sections through the petrous temporal bones and a sequence extending into the parotid glands. The imaging stigmata of a Bell's palsy may be demonstrated with a tuft on enhancement extending into the fundus of the IAM and marked enhancement within the anterior genu [63].

Cerebellopontine angle cistern, petrous temporal bone or parotid lesions may result in symptomatic facial nerve or vestibulocochlear nerve dysfunction and facial pain. As well as local facial pain from malignant lesions, it should be considered that trigeminal symptoms may occur the setting of facial nerve pathology due to central connections and due to perineural spread between facial and trigeminal nerve branches. This characteristic appearance may be seen within the auriculotemporal branch of the trigeminal nerve, extending to the facial nerve (Figure 10) and resulting in preauricular pain and facial nerve palsy [64]. Dysphagia, dysphonia or dysarthria may result from lesions of the caudal cranial nerves either in the brain stem, skull base or neck, and may be associated with facial pain. Again such pathology may be divided into inflammatory, vascular or neoplastic lesions. A lateral medullary infarct (Figure 11), associated with Wallenberg's syndrome may produce ipsolateral facial pain due to involvement of the spinal trigeminal tract or nucleus [65].

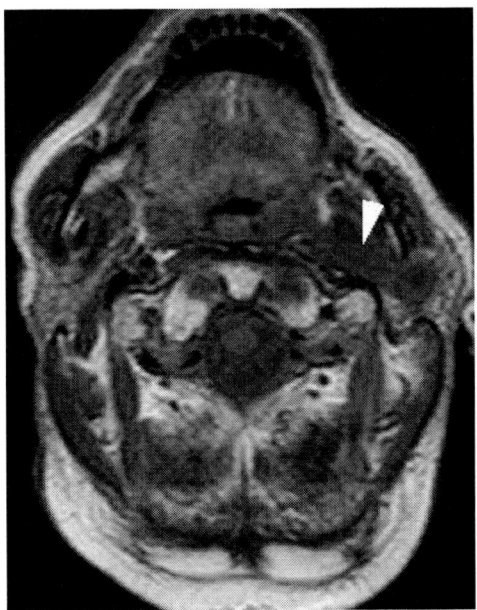

Figure 10. T1-w axial image reveals a band of intermediate signal tissue curving posterior to the mandibular ramus (white arrowhead). This corresponded to recurrent adenoid cystic carcinoma in a patient with auriculotemporal nerve perineural spread. Patient was experiencing facial pain and facial and trigeminal nerve dysfunction.

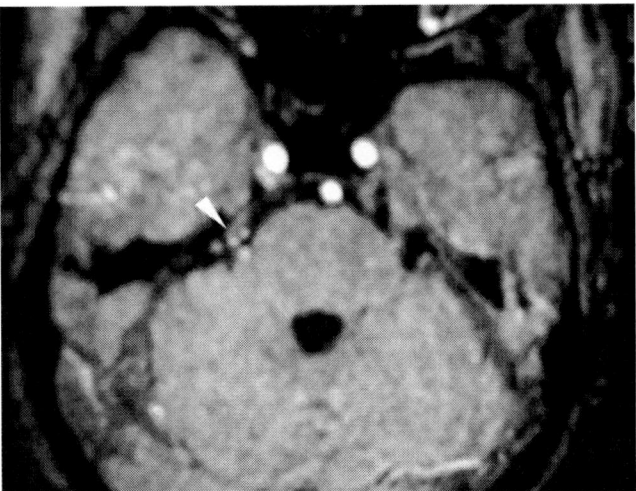

Figure 11. Diffusion weighted image illustrates increased signal in the left posterolateral medulla and left cerebellar hemisphere corresponding to PICA infarction. Patient developed trigeminal neuralgia in addition to other neurological symptoms.

The choice of imaging in this setting depends on whether the lesion is suspected to involve the brain or skull base, in which case MRI is necessary. If a lesion within the infrahyoid neck is felt possible, for instance in a patient with suspected distal vagal nerve palsy then CT is preferred.

## FACIAL PAIN REFERRED BY SHARED CRANIAL NERVE INNERVATION

Facial pain may be referred from structures distant to the site of the pain experienced. This phenomenon results from pain being referred along roots of shared innervation. Sensory innervation to the face, both superficially and deeply, is conveyed by trigeminal, facial, glossopharyngeal and vagus nerves, together with the upper cervical plexus. Therefore pathology in structures being innervated by the same cranial nerves may refer pain to structures innervated by the same nerve. For instance, facial pain is rarely a presenting feature of lung cancer due to shared innervation by the vagus nerve and since nociceptive signals within the cranial vault is largely transmitted by the ophthalmic division of the trigeminal nerve, a posterior fossa tumour may refer pain to the frontal region. Anatomical structures with trigeminal nerve and facial nerve sensory innervation will generally be covered in a scan volume, which includes, however nerves with a wider anatomical innervation such as the glossopharyngeal nerve (from the tympanic membrane to the hypopharynx) and the vagus nerve (from the bronchial tree to the external ear) may refer pain from outside the conventional scan volume (Figure 12).

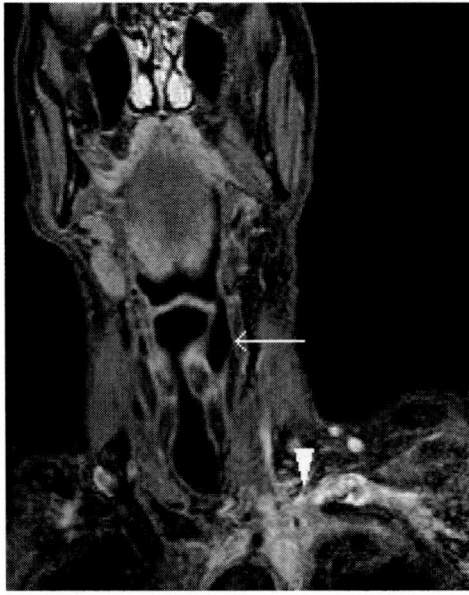

Figure 12. Coronal STIR image demonstrates a left apical carcinoma of the lung (Pancoast's tumour) (white arrowhead). There is also left distal vagal palsy as shown by the medially displaced left aryepiglottic fold (white arrow).

The clinical situation in which understanding of the sensory innovation is most important is that of otalgia [66]. Pain may be referred to the ear from the paranasal sinuses, oral cavity, pharynx, larynx parotid gland, thyroid gland, cervical spine, oesophagus and trachea, so the imaging net must be cast wide to include these sources if no local pathology is identified. A further pattern of referred pain results from central connections of the cranial nerves. All the sensory cranial nerves may converge on the trigeminal nucleus within the brain stem, so pathology within the end structures of these nerves may result in somatic distribution of pain within trigeminal nerve territory.

## OTHER NEUROLOGICAL FACIAL PAIN SYNDROMES

Although it is beyond the scope of this review to document all neurological facial pain syndromes, there are several in which the role of imaging have been discussed and explored. There are a group of trigeminal autonomic cephalgias (TACs), which are characterised by short lasting unilateral, predominantly fronto-orbital pain, with prominent ipsilateral cranial autonomic features [67]. This entity includes cluster headache, paroxysmal hemicrania, hemicrania continua and short lasting unilateral neuragiform headache attacks with conjunctival injection (SUNCT syndrome). They are rarely seen in the primary care setting but represented 6% of patients with facial pain referred to a neurological tertiary care centre in one study [68]. Secondary headache disorders that mimic TAC have been described secondary to infection, neoplastic and in particular vascular intracranial lesions [69]. Since secondary TAC is difficult to distinguish on clinical grounds, it has been suggested that MRI is performed, particularly if there is a new onset or atypical features [37, 67].

There are few studies of the diagnostic yield of imaging in the setting of pure midfacial pain. Imaging is generally employed to exclude local pathology in order to lend weight to a neurological diagnosis. One study of 127 patients with unilateral eye and facial pain and no neurological findings demonstrated only one responsible lesion [70]. Pure mid face pain is recognised by some authors to be a clinical presentation of chronic sinusitis. There is some data following up patients with endoscopy and CT negative facial pain in the settings of a nasal clinic which has shown that these patients did not respond well top medical or surgical treatment for sinonasal disease if they were CT and endoscopy negative [71, 72]. The majority of patients were treated for neurological diagnosis with clinical improvement. Hence there may be a role for low dose CT to select those patients with a negative scan, who will not benefit from sinonasal intervention and will be appropriate for neurological referral.

Many patients with negative imaging studies will be diagnosed with migraine. Migraneurs may experience pain primarily affecting the cheek, ear, nose and neck which are sometimes termed lower half headaches. Persistent idiopathic facial pain is a further classification of facial pain, which was previously named atypical facial pain. It is defined as a persistent facial pain, which does not have the classic characteristics of cranial neuralgia and is not attributed to another disorder [73, 74]. Since it is a diagnosis of exclusion, patients may undergo exhaustive imaging. Patient beliefs in this clinical setting are fundamental, and imaging may be used to reassure the patient that there is no serious pathology.

Carotidynia is defined as idiopathic neck pain with tenderness over the carotid bifurcation. It has been a diagnostic entity for decades, however this has more recently been removed as an entity within the clinical classification, partly due to the lack of histopathological and imaging correlation. Interestingly, there has been one recent imaging series, which has reported enhancing tissue around the carotid bifurcation in these patients on MRI [75].

## CONCLUSION

In conclusion, facial pain has been categorised in order to discuss the role of imaging. Trigeminal neuralgia is usually associated with neurovascular contact however it may be secondary to structural lesions along the trigeminal nerve pathway. There is a role for imaging in most cases, especially if the clinical presentation is atypical. Specific MRI sequences to depict neurovascular contact may be required if surgical intervention is considered. Facial pain with cranial nerve symptoms and signs, is frequently secondary to a structural abnormality and imaging is mandatory. Patients with pure facial pain are often imaged in order to exclude a structural cause prior to formulating a neurological diagnosis if the underlying cause is not clear after clinical examination. The imaging volume may need to cater for referred pain in problematic cases. Finally, other neurological facial pain syndromes are rarely secondary however imaging should be performed, particularly if there is an atypical presentation.

## REFERENCES

[1]   Zakrzewska JM. Facial pain:neurological and non-neurological. *J. Neurol. Neurosurg. Psych.* 2002; 72: 27-32.

[2]   Hapak L, Gordon A, Locker D et al. Differentiation between musculoligamentous, dentoalveolar, and neurologically based craniofacial pain with a diagnostic questionnaire. *J. Orofac. Pain* 1994; 8: 357-68.

[3]   Anonymous. Classification and diagnostic criteria for headache disorders, cranial neuralgias and facial pain. Headache Classification Committee of the International Headache Society Society. *Cephalgia* 1998; 8(Suppl7):1-96.

[4]   Katusic S, Beard CM, Bergstralh E et al. Incidence and clinical features of trigeminal neuralgia, Rochester, Minnesota, 1945-1984. *Ann. Neurol.* 1990; 27: 89-95.

[5]   Merskey H, Bogduk N. *Classification of chronic pain*. Descriptors of chronic pain syndromes and definitions of pain terms. Seattle: IASP Press, 1994:1.

[6]   Pearce JMS. Trigeminal neuralgia (Fothergill's disease) in the 17th and 18th centuries. *J. Neurol. Neurosurg. Psych.* 2003; 74: 1688.

[7]   Jannetta PJ. Arterial compression of the trigeminal nerve at the pons in patients with trigeminal neuralgia. *J. Neurosurg.* 1967; 26: 159-62.

[8]   Jannetta P. Microsurgical approach to the trigeminal nerve for tic doloreux. *Prog. Neurol. Surg.* 1976; 7: 180-200.

[9]   Devor M, Amir R, Rappaport ZH. Pathophysiology of trigeminal neuralgia:The ignition hypothesis. *Clin. J. Pain* 2002; 18: 4-13.

[10]  Cheshire WP. Trigeminal neuralgia: diagnosis and treatment. *Curr. Neurol. Neurosci. Rep.* 2005; 5: 79-85.

[11] Barker FG, Jannetta PJ, Bissonette DJ et al. The long term outcome of microvascular decompression for trigeminal neuralgia. *N. Eng. J. Med.* 1996; 334: 1077-1083.

[12] Anderson VC, Berryhill PC, Sandquist MA et al. High resolution three-dimensional magnetic resonance angiography and three-dimensional spoiled gradient-recalled imaging in the evaluation of neurovascular compression in patients with trigeminal neuralgia: a double-blind pilot syudy. *Neurosurgery* 2006; 58: 666-73.

[13] Benes L, Shiratori K, Gurschi M et al. Is preoperative high-resolution magnetic resonance imaging accurate in predicting neurovascular compression in patients with trigeminal neuralgia? A single-blind study. *Neurosurg. Rev.* 2005; 28: 131-6.

[14] Boecher-Schwarz HG, Bruehl K, Kessel G et al. Sensitivity and specificity of MRA in the diagnosis of neurovascular compression in patients with trigeminal neuralgia. A correlation of MRA and surgical findings. *Neuroradiology* 1998; 40: 88-95.

[15] Jawahar A, Kondziolka D, Kanal E et al. Imaging the trigeminal nerve and pons before and after surgical intervention for trigeminal neuralgia. *Neurosurgery* 2001; 48: 101-7.

[16] Korogi Y, Nagahiro S, Du C eta l. Evaluation of vascular compression in trigeminal neuralgia by 3D time-of flight MRA. *J. Comput. Assist. Tomogr.* 1995; 19: 879-84.

[17] Majoie CB, Hulsmans FJ, Verbeeten BJ et al. Trigeminal neuralgia: comparison of two MR imaging techniques in the diagnosis of neurovascular contact. *Radiology* 1997; 204: 455-60.

[18] Masur H, Papke K, Bongartz G et al. The significance of three-dimensional MR-defined neurovascular compression for the pathogenesis of trigeminal neuralgia. *J. Neurol.* 1995; 242: 93-8.

[19] Meaney JF, Eldridge PR, Dunn LT et al. Demonstration of neurovascular compression in trigeminal neuralgia with magnetic resonance imaging. Comparison with surgical findings in 52 consecutive operative cases. *J. Neeurosurg.* 1995; 83: 799-805.

[20] Meaney JF, Miles JB, Nixon TE et al. Vascular contact with the fifth cranial nerve at the pons in patients with trigeminal neuralgia: detection with 3D FISP imaging. *Am. J. Roentgenol.* 1994; 163: 1447-52.

[21] Tanaka T, Morimoto Y, Shiiba S et al. Utility of magnetic resonance cisternography using three-dimensional fast asymmetric spin-echo sequences with multiplanar reconstruction: The evaluation of sites of neurovascular compression of the trigeminal nerve. *Oral Surg. Oral Med. Oral Pathol. Oral Radiol. Endod.* 2005; 100:215-25.

[22] Yamaki T, Kobayashi S, Hirschberg H et al. Preoperative assessment of trigeminal neuralgia and hemifacial spasm using constructive interference in steady state-three-dimensional Fourier transformation magnetic resonance imaging. *Neurol. Med. Chir.* (Tokyo) 200; 40: 545-6.

[23] Yoshino N, Akimoto H, Yamada I. Trigeminal neuralgia: Evaluation of neuralgic manifestation and site of neurovascular compression with 3D CISS MR imaging and MR angiography. *Radiology* 2003; 228: 539-545.

[24] Brisman R, Khandji AG, Mooij RB. Trigeminal nerve-blood vessel relationship as revealed by high resolution magnetic resonance imaging and its effect on pain relief after gamma knife radiosurgery for trigeminal neuralgia. *Neurosurgery* 2002; 50: 1261-6.

[25] Cheuk AV, Chin LS, Petit JH. Gamma knife surgery for trigeminal neuralgia: outcome, imaging, and brainstem correlates. *Int. J. Radiat. Oncol. Biol. Phys.* 2004; 60: 537-41.

[26] Erbay SH, Bhadelia RA, Riesenburger R et al. Association between neurovascular contact on MRI and response to gamma knife radiosurgery in trigeminal neuralgia. *Neuroradiology* 2006; 48: 26-30.

[27] Peters G, Nurmikko TJ. Peripheral and gasserian ganglion-level procedures for the treatment of trigeminal neuralgia. *Clin. J. Pain* 2002; 18: 28-34.

[28] Sindou M, Howeidy T, Acevedo G. Anatomical observations during microvascular decompression for idiopathic trigeminal neuralgia (with correlations between topography of pain and site of the neurovascular conflict). Prospective study in a series of 579 patients. *Acta Neurochir.* (Wein) 2002; 144: 1-12.

[29] Hardy DG, Rhoto AL. Microsurgical relationship of the superior cerebellar artery and the trigeminal nerve. *J. Neurosurg.* 1978; 49: 669-78.

[30] Haines SJ, Jannetta PJ, Zorub DS. Microvascular relations of the trigeminal nerve. *J. Neurosurg.* 1980; 52: 381-6.

[31] Klun B, Prestor B. Microvascular relations of the trigeminal nerve: an anatomical: an anatomical study. *Neurosurgery* 1986; 19: 535-8.

[32] Lang J. Clinical anatomy of the posterior cranial fossa and its foramina. Stuttgart: *Georg Thieme Verlag*, 1991:82.

[33] Lang E, Naraghi R, Tanrikulu L. Neurovascular relationship at the trigeminal root entry zone in persistent idiopathic facial pain: findings from MRI 3D visualisation. *J. Neurol. Neurosurg. Psych.* 2005; 76: 1506-1509.

[34] Macfarlane TV, Blinkhorn AS, Davies RM et al. Orofacial pain: just another chronic pain? Results from a population based survey. *Pain* 2002; 99: 453-8.

[35] Jamjoom AB, Jamjoon ZAB, Al-Fehaily M et al. Trigeminal neuralgia related to cerebellopontine angle tumors. *Neurosurg. Rev.* 1996; 19: 237-241.

[36] Puca A, Meglio M, Tamburrini G et al. Trigeminal involvement in intracranial tumors. Anatomical and clinical observations in 73 patients. *Acta Neurochir* (Wien) 1993; 125: 47-51.

[37] Siccoli MM, Basetti CL, Sandor PS. Facial pain: clinical differential diagnosis. *Lancet Neurol.* 2006; 5: 257-67.

[38] Cheng TMW, Cascino TL, Onofrio BM. Comprehensive study of diagnosis and treatment of trigeminal neuralgia secondary to tumors. *Neurology* 1993; 2298-302.

[39] Matthies C, Samii M. Management of 1000 vestibular schwannomas (acoustic neuromas): clinical presentation. *Neurosurgery* 1997; 40: 1-9.

[40] Ogleznev KY, Grigoryan YA, Slavin KV. Parapontine epidermoid tumors presenting as trigeminal neuralgias. Anatomical findings and operative results. *Acta Neurochir* (Wien) 1991; 110: 116-119.

[41] Rubin G, Scienza R, Pasqualin A et al. Craniocerebral epidermoids and demoids. A review of 44 cases. *Acta Neurochir* (Wien) 1989; 97: 1-16.

[42] Ginsberg LE, DeMonte F. Imaging of perineural tumor spread from palatal carcinoma. *AJNR Am. J. Neuroradiol.* 1998; 19: 1417-1422.

[43] Cheng JS, Sanchez-Mejia RO, Limbo M. Management of medically refractory trigeminal neuralgia in patients with multiple sclerosis. *Neurosurg. Focus 2005*; 18: 13.

[44] Hooge JP, Redekop WK. Trigeminal neuralgia in multiple sclerosis. *Neurology* 1995; 45: 1294-6.

[45] Iizuka O, Hosokai Y, Mori E. Trigeminal neuralgia due to pontine infarction. *Neurology* 2006; 66: 48.

[46] Ito M, Sonokawa T, Mishina H et al. Dural arteriovenous malformation manifesting as tic douloureux. *Surg. Neurol.* 1996; 45: 370-75.
[47] Rushton JG, Stevens JC, Miller RH. Glossopharyngeal (vagoglossopharyngeal) neuralgia: a study of 217 cases. *Arch. Neurol.* 1981; 38: 201-5.
[48] Patel A, Kassam A, Horowitz M. Microvascular decompression in the management of glossopharyngeal neuralgia: analysis of 217 cases. *Neurosurgery* 2002; 50: 705-10.
[49] Montalbetti L, Ferrandi D, Pergami P et al. Elongated styloid process and Eagle's syndrome. *Cephalgia* 1995; 15: 80-93.
[50] Optic Neuritis Study Group. The clinical profile of optic neuritis: experience of the Optic Neuritis Treatment Trial. *Arch. Ophthalmol.* 1991; 109: 1673-78.
[51] Optic Neuritis Study Group. The 5-year risk of MS after optic neuritis:experience of the Optic Neuritis Treatment Trial. *Neurology* 1997; 49: 1404-13.
[52] Hickman SJ, Miszkiel KA, Plant GT, Miller DH. The optic nerve sheath on MRI in acute optic neuritis. *Neuroradiology* 2005; 47: 51-5.
[53] Wong GK, Boet R, Poon WS et al. A review of isolated third nerve palsy without subarachnoid hemorrhage using computed tomographic angiography as the first line investigation. *Clin. Neurol. Neurosurg.* 2004; 107: 27-31.
[54] Schievink WI. Spontaneous dissection of the carotid and vertebral arteries. *N. Engl. J. Med.* 2001; 344: 898-906.
[55] Majoie CBLM, Hilsmans F-J H, Castelijns JA et al. Symptoms and signs related to the trigeminal nerve: Diagnostic yield of MR imaging. *Radiology* 1998; 209: 557-562.
[56] Lossos A, Siegal T. Numb chin syndrome in cancer patients: etiology, response to treatment and prognostic significance. *Neurology* 1992; 42: 1181-1184.
[57] Hutchins LG, Harnsberger HR, Hardin CW. *The radiologic assessment of trigeminal neuropathy.*
[58] Lecky BRF, Hughes RAC, Murray NMF. Trigeminal sensory neuropathy: a study of 22 cases. *Brain* 1987; 110:1463-1485.
[59] Gilden DH, Kleinschmidt-DeMasters NK, LaGuardia JJ et al. Neurologic complications of the reactivation of varicella-zoster virus. *N. Engl. J. Med.* 2000; 342: 635-45.
[60] Quisling S, Shah VA, Lee HK et al. Magnetic resonance imaging of third cranial nerve palsy and trigeminal sensory loss caused by herpes zoster. *J. Neuro-ophthalmol.* 2006; 26: 47-48.
[61] Gilden DH. Bell's Palsy. *N. Engl. J. Med.* 2004; 351: 1323-31.
[62] Kress B, Griesbeck F, Stippich C et al. Bell palsy: quantitative analysis of MR imaging data as a method of predicting outcome. *Radiology* 2004; 230: 504-9.
[63] Tien R, Dillon WP, Jackler RK. Contrast-enhanced MR imaging of the facial nerve in 11 patients with Bell's palsy. *AJNR Am. J. Neuroradiol.* 1990; 11: 735-41.
[64] Schmalfuss IM, Tart RP, Mukherji et al. Perineural tumor spread along the auriculotemporal nerve. *AJNR Am. J. Neuroradiol.* 2002; 23: 303-11.
[65] Fitzek S, Baumgartner U, Fitzek C et al. Mechanisms and predictors of chronic facial pain in lateral medullary infarction. *Ann. Neurol.* 2001; 49: 493-500.
[66] Weissman JL. A pain in the ear: The radiology of otalgia. *AJNR Am. J. Neuroradiol.* 1997; 18: 1641-1651.
[67] Matharu MS, Goadsby PJ. Trigeminal autonomic cephalgias. *J. Neurol. Neurosurg. Psych.* 2002; 72: 19-26.

[68] Zebenholzer K, Wober C, Vigl M et al. Facial pain in a neurological tertiary care centre-evaluation of the International Classification of Headache Disorders. *Cephalgia* 2005; 25: 689-99.

[69] Galende AV, Camacho A, Gomez-Escalonilla C et al. Lateral medullary infarction secondary to vertebral artery dissection presenting as a trigeminal autonomic cephalgia. *Headache* 2004; 44: 70-74.

[70] Harooni H, Golnik KC, Geddie B. Diagnostic yield for neuroimaging in patients with unilateral eye or facial pain. *Can. J. Ophthalmol.* 2005; 40: 759-63.

[71] Paulson EP, Graham SM. Neurologic diagnosis and treatment in patients with computed tomography and nasal endoscopy negative facial pain. *Laryngoscope* 2004; 114: 1992-6.

[72] West B, Jones NS. Endoscopy-negative, computed tomography-negative facial pain in a nasal clinic. *Laryngoscope* 2001; 111: 581-6.

[73] The International Classification of Headache Disorders, second edn. *Cephalgia* 2004: 24(suppl 1): 9-160.

[74] Pfaffenrath V, Rath M, Pollmann W et al. Atypical facial pain: application of the HIS criteria in a clinical sample. *Cephalgia* 1993; 13(Suppl 12): 84-8.

[75] Burton BS, Syms MJ, Petermann GW, Burgess LP. MR imaging of patients with carotidynia. *AJNR Am. J. Neuroradiol.* 2000; 21: 766-9.

*Chapter 9*

# SPECT AND PSYCHIATRY – AN INSIGHT INTO FUNCTIONAL IMAGING OF MAJOR DEPRESSION AND POST-TRAUMATIC STRESS DISORDER

## *Marco Pagani*[*,1,2] *and Ann Gardner*[3]

[1]Institute of Cognitive Sciences and Technology, CNR, Rome, Italy
[2]Section for Nuclear Medicine, Department of Hospital Physics,
Karolinska Hospital, Stockholm, Sweden
[3] Karolinska Institutet, Department of Clinical Neuroscience,
Section of Psychiatry, Karolinska University Hospital Huddinge,
Stockholm, Sweden

**Keywords.** $^{99m}$Tc-d,l-hexamethylpropyleneamine oxime, Single Photon Emission Computed Tomography, Major Depressive Disorder, Post Traumatic Stress Disorder.

## ABBREVIATIONS

| | |
|---|---|
| 2-D | two-dimensional |
| 3-D | three-dimensional |
| AD | Alzheimer Disease |
| ANOVA | analysis of variance |
| ATP | adenosine triphosphate |
| CBA | computerized brain atlas |
| CBF | cerebral blood flow |
| DA | discriminant analysis |
| FLD | frontal lobe dementia |
| fMRI | functional magnetic resonance imaging |

---

[*] Corresponding Author: Marco Pagani MD PhD, Institute of Cognitive Sciences and Technologies, CNR. Via S.Martino della Battaglia 44. 00185, Rome, Italy. Tel: +39-06-44595321. Fax: +39-06-44595243. e-mail: marco.pagani@istc.cnr.it

| | |
|---|---|
| GSH | reduced glutathione |
| MDD | major depressive disorder; also referred to as "depression" in the text |
| mtDNA | mitochondrial DNA |
| PCA | principal component analysis |
| PD | Parkinson Disease |
| PET | positron emission tomography |
| rCBF | regional cerebral blood flow |
| rCMRGlu | regional cerebral metabolic rates of glucose |
| rCBV | regional cerebral blood volume |
| ROC | receiver operating characteristic |
| ROI | region of interest |
| SERT | serotonin transporter |
| SPECT | single photon emission computed tomography |
| SPM | statistical parametric mapping |
| VOI | volume of interest |

# INTRODUCTION

During the last 30 years functional brain imaging has been increasingly applied to psychiatric disorders. In the field of neurodegenerative disorders, single photon emission computed tomograpghy (SPECT) and positron emission tomography (PET) allow nowadays for the identification of mild-to-severe forms of Alzheimer's Diseases (AD), Frontal Lobe Dementia (FLD) and Parkinson Disease (PD) with a sensitivity and specificity approaching 80-90%. Such high accuracy has not been realized for other forms of dementia (i.e. vascular dementias and Dementia with Lewy's Bodies), for mild cognitive impairment, for schizophrenia, for post-traumatic stress disorder or for all forms of depression in which the links between the findings of functional brain imaging studies and the neural substrates of such disorders have not been clearly established yet.

In this scenario, the importance of the combination of newly developed imaging and statistical techniques in the assessment of psychiatric disorders steadily increases. The neuropsychiatric and behavioural abnormalities are often a source of considerable patient and caregiver distress, whilst proper diagnosis also contributes to the decision of the level of care of these patients resulting in early diagnosis and better patient management with considerable savings for the community.

Optimised nuclear medicine techniques and algorithms for functional brain imaging could now be implemented in the clinical management of psychiatric patients for finer discrimination of cerebral blood flow (CBF) or metabolic changes. Such changes have for long time been neglected due to the fact that the quality of both functional images and image analysis was not sufficient to identify the sometimes small functional regional variations occurring in psychiatric diseases.

In this respect it is of utmost importance to highlight that in recent decades a general consensus on the CBF distribution pattern in AD and FLD has been built and that clinical diagnosis is now routinely based on the visual identification of reduced radiopharmaceutical uptake in the temporo-parietal cortex. Such consensus is still lacking for psychiatric disorders in which a specific CBF distribution pattern has not yet been identified.

The present review concentrates on state-of-the-art technologies and on recent investigations dealing with SPECT and psychiatry. As for neurodegenerative diseases (i.e. AD, FLD and PD) extensive reviews are widely available [1, 2, 3] and schizophrenia has recently been reviewed in this journal [4]. We will introduce and briefly comment on the most recent methodological advancements and statistical techniques. We will also discuss two psychiatric disorders in which functional neuroimaging has recently played a substantial role in identifying the neurobiological changes and, in perspective, in improving the clinical diagnosis.

## METHODOLOGICAL CONSIDERATIONS

### PET and SPECT

Functional mapping of the brain by PET is considered superior to SPECT due to the applicability of positron-emitting "bioisotopes", oxygen-15, nitrogen-13 and carbon-11, to the better spatial resolution and to the possibility of performing quantitative assessment. Nevertheless, its availability is restricted by the heavy on-site capital investment required to build an in-house cyclotron facility, even if nowadays several centres perform the studies with commercially available positron emitters. Due to this, the number of PET cameras remains relatively small, as compared to the numerous SPECT facilities, making SPECT the most commonly used technique especially in routine examinations.

However, it is worth noting that substantial differences exist between SPECT systems in comparison with PET.

SPECT imaging utilizing the radiotracers $^{133}$Xenon ($^{133}$Xe), $N$-isopropyl-$p[^{123}$I]-iodoamphetamine ($^{123}$IMP), $^{99m}$Tc - $d,l$ – hexamethylpropylene amine oxime ($^{99m}$Tc – HMPAO, $^{99m}$Tc-Exametazime), and $^{99m}$Tc – ethyl cysteinate dimer ($^{99m}$Tc – bicisate, $^{99m}$Tc – ECD) is generally considered to reflect regional cerebral blood flow (rCBF), and neuronal activity. $^{99m}$Tc-HMPAO is a commonly used tracer at SPECT investigations and in this review description of tracer properties and findings in psychiatric disorders will mostly refer to this radiopharmaceutical in order to present comparable data.

Using SPECT and $^{99m}$Tc-HMPAO to assess CBF, absolute quantitation of CBF is difficult since the tracer consists of lipophilic and hydrophilic components which cannot be separately monitored by blood sampling. In addition, $^{99m}$Tc-HMPAO uptake has been shown to be flow-dependent [5], systematically underestimating high-flow regions. In SPECT, relatively low spatial resolution causes a larger partial volume effect as compared to PET. The optimal system spatial resolution of present high-resolution PET cameras is of the order of 3-5 mm, while that of advanced SPECT cameras is of the order of 7-9 mm and in SPECT, correction for attenuation is mostly performed using some approximate method while scatter correction is generally not implemented. For such reasons it is sometimes difficult to identify by SPECT relatively small brain structures like amygdala, involved in the pathophysiology of PTSD and other anxiety disorders.

On the other hand, SPECT has the unique advantage on PET to allow for experiments to be performed in ideal psychological conditions in a quite environment outside the camera gantry. In fact, due to the characteristics of SPECT and $^{99m}$Tc, image acquisition can be started up to some hours after injection still representing the radiopharmaceuticals brain distribution at the moment of injection.

A dynamic coupling between CBF as assessed by SPECT and brain metabolism as assessed by regional cerebral metabolic rates of glucose (rCMRGlu) as assessed by PET, has generally been assumed.

A reliable diagnosis is achieved by identifying those structures, irrespective of the size, in which the modification of CBF or metabolism deviate from normality resulting in significant changes as compared to a reference database. In fact, the diagnostic value of a functional brain image in psychiatric studies is increased if the patient's scan can be compared to an average scan obtained from a group of control subjects.

The capability of SPECT to detect fine functional and pathological changes is strictly dependent on the implementation of high resolution cameras, on sophisticated and dedicated software able to automatically standardize the brain space and data across scans and on the use of advanced statistical methods.

## Visual Evaluation and Manual or Semi-Manual Outlining

The assessment of CBF patterns in various brain disorders by SPECT or PET have in the past mainly been carried out either by visual evaluation [6-9] or by outlining the regions of interest (ROIs) in a manual or semi-automatic mode [10-13]. Results were obtained by computing ratios between target and reference regions.

Such methods are time-consuming and poorly reproducible and may suffer from excessive operator's influence in the choice of the ROIs due to the variable shape of human brains and lack of spatial normalization, thus resulting in anatomical in-homogeneous brain samples among subjects.

## Multidimensional Semi-Automatic Approaches

In the recent past a three-dimensional (3-D) semi-automatic approach for the identification of brain regions and CBF changes in a single individual as compared to a population of patients and normal controls has been implemented.

The 3-D analysis, as compared to conventional 2-D data representation on transversal slices, is less dependent on errors in outlining position. Furthermore, analysis of volumes of interest (VOIs) reduces the variance due to counting statistics since the number of voxels in a functional region (VOI) is larger than a number of pixels in a 2-D ROI. In 3-D analysis, the inclusion of the white matter makes the sample more representative to global changes and VOIs can be positioned on both anatomical and functional regions improving the physiological significance of the analysis. White matter is an important part of the neuronal system and it is affected by psychiatric disorders to the same extent as grey matter, its perfusion being 2.0-3.5 times lower than the perfusion of grey matter [14-15].

By carefully standardizing each scan to an age-related database of control subjects it is possible, by means of subtraction images and/or statistical comparisons, to precisely identify regions with abnormal flow.

## Standardization Software

In recent years, several 3D digitized spatial standardization software has been proposed and some have been extensively used both in research and clinical investigations [16-17]. Most share similar principles and can be classified into two categories: the voxel-based (i.e. Statistical Parametric Mapping (SPM), NeuroStat (3D-SSP) and Brain Registration and Analysis Software Suite, BRASS) and those based on neuroanatomy (i.e. Computerized Brain Atlas, CBA). The advantage of these techniques is the possibility to fully exploit the knowledge of the rCBF pattern as assessed in a group of normal subjects by using it as a reference for studies of patients. Control subjects are grouped according to their age and a reference image containing the CBF information for all subjects is created. To highlight the possible pathology in most of the cases, the patient image is subtracted from the reference one. However, in psychiatric disorders, the alteration sometimes results in an increased rCBF. In this case, subtracting the reference image from the patient image will highlight the changes.

## Gamma Cameras

Most SPECT systems today are based upon rotating gamma cameras. A gamma camera is a position sensitive device that allows visualization of the distribution of gamma emitting radiotracers. A gamma camera is built up by a collimator, a scintillation detector, light detecting photomultiplier tubes (PM-tubes) and electronics for signal processing. Modern gamma camera systems are operating on-line to a computer for additional signal processing, image processing, tomographic reconstruction and image display. Due to the complex brain anatomy, tomographic examination is a pre-requisite for tracer distribution studies at SPECT when superimposed anatomo-functional scans are needed. The resulting contrast-enhancing effect of the tomographic co-registration technique is of fundamental importance as in psychiatry, the differences between the normal and pathological distribution in various brain regions can be rather small [18].

A multi-hole collimator constitutes the major imaging-forming part of a gamma camera. A parallel-hole collimator accepts only those photons that are incident in perpendicular direction to the planar detector surface thus decreasing the sensitivity but increasing the spatial resolution. Those photons which are allowed to pass through the collimator will enter into the scintillation detector. Visible light flashes are emitted from the local site where the photon is absorbed in the 9.5 mm thin NaI(Tl)-scintillator. The light is detected by closely packed PM-tubes on the backside of the scintillator. The amplitude of the fast output electronic signals ($< 1\mu s$) from the PM-tubes depends on the fraction of incident light that hits each individual PM-tube from a photon absorption event. For each event, the signals from the whole PM-tube's network are weighed together to give the two-dimensional (2-D) position coordinates and a measure of the energy of the photon absorbed. The energy signal is approximately proportional to the photon energy absorbed in the scintillator. It is thus possible to reject unwanted photons with origins from "Compton scattering" in the subject or in the detector by selecting an electronic "energy window" corresponding to the energy signals of primary, non-scattered photons.

The image quality of a gamma camera system is generally characterized in terms of spatial resolution, system sensitivity and energy resolution. "Full Width at Half Maximum" (FWHM) of count density along a profile across a point- or a line-activity source is a usual expression of the spatial resolution of the system and of the imaging sharpness of the system. The sensitivity (i.e. the number of detected photons per unit of activity) of a SPECT system is generally very low and of the order of $10^{-4}$ detected events per emitted photon. The sensitivity and spatial resolution are restricted by the properties of the collimator that is used. High sensitivity is counterbalanced by low spatial resolution.

Modern SPECT cameras are often designed with more than one camera head to improve sensitivity, with a three-head system giving a three-fold sensitivity increase as compared to a single-head camera. The energy resolution is of great importance for the ability to reject "falsely-positioned" scatter events. Images representing 2-D projection of a 3-D distribution are acquired from many different angles around the subject during a SPECT acquisition. The 3-D radioactivity distribution can be reconstructed by an algorithm given the 2-D projections.

## Normalization and the Reference Region

Normalization of the measured tracer uptake is necessary in the semi-quantification of SPECT data. The normalization procedure rates each pixel or voxel (the smallest units of a 2-D picture or a 3-D volume, respectively) as a proportion of the average of those pixels or voxels that are chosen as reference. Normalization of the cortical activity to the global mean, cerebellum, or one or more of the central structures, is unsuitable in many situations. Pathological processes with decreased or increased regional distribution of the tracer, as well as the spatial resolution of the SPECT system, may affect a reference region. A normalization procedure where a proportion of the voxels with the highest values are used as reference may be less sensitive in psychiatric disorders in which the pathological status is reflected by an increased tracer retention[18].

## Semi-Quantitative Versus Quantitative Measures

The SPECT technique is a semi-quantitative method, reflecting the distribution of the radiotracer and results are usually expressed in qualitative values, i.e., the reading physician bases the image interpretation on subjective evaluation, with the diagnostic interpretation being performed by visually comparing the intensity of the tracer distribution between the hemispheres, or by matching the signal intensity in a defined region to the contra-lateral region or to a reference region. At visual interpretation it is always uncertain whether an asymmetry is due to a decrease on one side or to an increase on the contra-lateral side. A certain degree of asymmetry must also be accepted as normal. A mere qualitative analysis of data may not reveal early or bilateral changes of tracer distribution. Digital data are necessary

## $^{99m}$Tc-HMPAO Properties

Intracellular trapping of lipophilic d,l - $^{99m}$Tc-HMPAO and its conversion to hydrophilic form has been considered to be the basis of retention of the tracer [19, 20]. The conversion has been related to the cellular content of reduced glutathione (GSH), a non-protein thiol, which is present at high intracellular concentrations [21, 22]. The tissue uptake for the d,l - $^{99m}$Tc-HMPAO - isomer in the brain, heart and liver, where conversion to the retainable (non-diffusible) form is large enough, has been suggested to reflect blood flow [22].

Other phenomena than intracellular GSH content may contribute to the retention of $^{99m}$Tc-HMPAO in the brain [23, 24, 25]. Decrease of intracellular $^{99m}$Tc-HMPAO has been linked to increased extracellular $^{99m}$Tc-HMPAO conversion to the hydrophilic form due to a hyper-reduced state in the extracellular space, with a decrease of the remaining extracellular hydrophilic $^{99m}$Tc-HMPAO fraction for cell entry and intracellular retention. Intracellular retention of $^{99m}$Tc-HMPAO may be as dependent upon the intracellular redox state as well as upon GSH content [23]. In a study of the $^{99m}$Tc-HMPAO metabolism in brain homogenate, the mitochondrial fraction showed a two-fold higher $^{99m}$Tc-HMPAO activity compared to the cytosol fraction. This finding may reflect an involvement of non-protein thiols with higher reductive activity than glutathione in mitochondria, with ensuing increased mitochondrial $^{99m}$Tc-HMPAO retention. Hyperfixation of $^{99m}$Tc-HMPAO in the brain may indicate damaged mitochondria [25].

## Statistical Considerations

Functional imaging is based on complex technology and statistical techniques. The required variables may be significantly more than the investigated subjects, which confers severe limitations to the conclusions that can be drawn from studies with small samples. Most studies in brain research have been performed by analysing a small selected number of regions according to a predetermined working hypothesis, with a selection of regions based on previous data in the literature. Identification of the neural substrates involved in psychiatric disorders will therefore necessitate a number of studies from various centres before observations can be considered as confirmed.

Standard statistical methods might not be sufficient to analyse the large amount of data generated by whole brain functional imaging and different pre-processing algorithms therefore data analysis models may be of use [26, 27, 28, 29].

## Univariate Analysis

In univariate analysis there is only one variable under consideration. It can be independent or dependent as in the case of the same subject measured at two different times. In both cases it is possible to describe the data in terms of mean and variance (the two parameters of the normal distribution). After testing the two means, possible significant differences need to be explained. The standard approach is to assume that the difference is due to an experimental effect and sources of variance are under control. However, this is not so obvious in neurofunctional studies in which many sources of variance are present. Hence, if we want to study the relationships among those sources, multivariate analysis has to be implemented.

The t-test for dependent samples is the most commonly used method to evaluate the differences in means between two groups of observation made on the same sample of subjects who were tested twice. When groups of observation are made on different subjects a t-test for independent samples is used. One-way ANOVA is performed when groups are three or more. In such cases nothing can be done about the variation due to individual differences since it is not possible to identify, or subtract, such differences. This is why the t-test for independent samples is always less sensitive.

In brain imaging, univariate analysis is typically performed by SPM [30]. SPM and is the predominantly used worldwide voxel-based standardization software in brain imaging for between- and within-subject CBF comparisons. Images are spatially standardized into a common space and smoothed. Parametric statistical models are summed, at each voxel, using the General Linear Model to describe the variability in the data. Hypotheses expressed in terms of the model parameters are assessed at each voxel with univariate statistics. This results in an image whose voxel values are statistics, producing t-statistical maps of significant changes in distribution and basing the output on the analyses of clusters of voxels. Such analysis should take into account the statistical threshold as well as the size of the cluster in relationship to the implemented methodology: the higher the spatial resolution of the camera the smaller the size of cluster of voxels for statistical significance.

## Multivariate Analysis

Multivariate statistics provide simultaneous analysis of multiple independent (i.e. sex, disease) and dependent (i.e. Hemispheres, VOI) variables in order to determine the relationship among them. Such statistical approach also introduces regional analyses based on the assumption that correlated patterns exist among different brain regions and such relationships affect reciprocally the investigated variable. Variables may be correlated with each other, and their statistical dependence is often taken into account when analyzing such data. In fact, the consideration of statistical dependence and intercorrelations between variables make multivariate analysis somewhat different in approach and considerably more complex than the corresponding univariate analysis in which there is only one variable under consideration. In the multivariate perspective each voxel is considered conjointly with explicit reference to the interactions among brain regions rendering it particularly appropriate for brain studies and providing a complementary characterization of CBF patterns.

Multivariate analysis requires the number of observations (scans) to be greater than the number of components of the multivariate observation (variables, i.e. voxels). In neuroimaging techniques (in which the raw images contain an extremely high number of voxels) the number of variables need, therefore, to be reduced by using ROIs, VOIs or factorial groupings.

It takes into account the statistical inference about the response of the entire brain without regional specificity. If interactions are present one can move from an "omnibus" effect to regional changes with the limitation of the sample size (ROI/VOI/factor).

## ANOVA, MANOVA, DA, ROC AND OTHER STATISTICAL METHODS

Multivariate statistics include Analysis of Variance (ANOVA), Discriminant Analysis (DA) and Receiver Operating Characteristic (ROC) analysis, K-means clustering and Principal Component Analysis (PCA). ANOVA (and MANOVA, multivariate ANOVA) is used when the design involves one or more categorical independent variables (i.e. groups) and two or more continuous dependent variables (i.e. CBF or metabolism). As well as identifying whether changes in the independent variables have a significant effect on the dependent variables, the technique also seeks to identify the interactions among the independent variables and the association between dependent variables.

DA is performed at group level to estimate the relationship between groupings performed according to the tested methodology and a gold standard (i.e. SPECT/PET diagnosis vs previous clinical diagnosis). The main objective is to construct rules for assigning future observations to one of the groups in order to minimize the probability of misclassification. If variables are effective for a set of data, the classification table of correct and incorrect estimates will yield a high percentage correct.

The ability of a test to discriminate diseased cases from normal cases could also be visually evaluated using ROC curve analysis [31, 32]. A ROC curve is simply a plot of the true positive (sensitivity) rate against the false positive rate for the different possible cut-off points of a diagnostic test. As would be expected, achieving higher detection performance generally results in an increase in incidents of false alarms: any increase in sensitivity will be accompanied by a decrease in specificity.

K-means clustering is implemented to create k groups of individuals based on raw data. It splits a set of objects into a selected number of groups by maximizing between variation relative to within variation. The procedure iterates through the data until assigning cases to a specified number of non-overlapping clusters. Chi-square is used to test the distribution differences between the obtained clusters and the clinical diagnosis and type.

PCA is a data driven technique (i.e., there is no a-priori model or hypothesis) that transforms a number of (possibly) correlated variables into a (smaller) number of not-correlated factors, called principal components. PCA is totally data-led and is independent by any model or a-priori hypothesis. It does not create effects that are not present in the data, nor loses information. The first principal component accounts for the highest percentage of the variability in the data and each of the following components account for a portion of the remaining variability in a descending scale. This statistical approach introduces regional analyses based on the assumption that correlated patterns exist among different brain regions and such relationships affect reciprocally the rCBF or the metabolism. In PCA each component is orthogonal and functionally not correlated to the remaining ones. PCA is of particular relevance in the study of functional connectivity.

## Functional Connectivity

Functional connectivity implies that pool activities of brain areas change together and regions share a significant number of neurons whose dynamic interactions occur at the same time. Correlated areas will have correlated perfusion and neuronal activity. Functional connectivity is simply a statement about the observed correlations and characterizes distributed brain systems.

The functional role played by any component (neuron) of a connected system (brain) is largely defined by its connections. Extrinsic connections between cortical areas are not continuous but occur in patches or clusters (functional segregation, in which cells with common functional properties are thought to be grouped together).

On the other hand functional integration is mediated by the interactions between functionally segregated areas resulting in a general functional connectivity effect on the brain. Functional connectivity characterizes distributed brain systems and implies "model-free" temporal correlations between neurophysiological events: correlated areas will have correlated perfusion and neuronal activity.

The issues related to functional segregation are generally investigated by means of univariate analysis while functional integration is better analyzed by multivariate analysis. SPM (univariate analysis) is typically predicated by functional segregation and analyses regionally specific aspects of functional organization. PCA and multivariate analysis are inspired by functional integration mediated by anatomical, functional and effective connections that form the basis for characterizing patterns of correlations and describe distributed changes in terms of systems (see Figure 1).

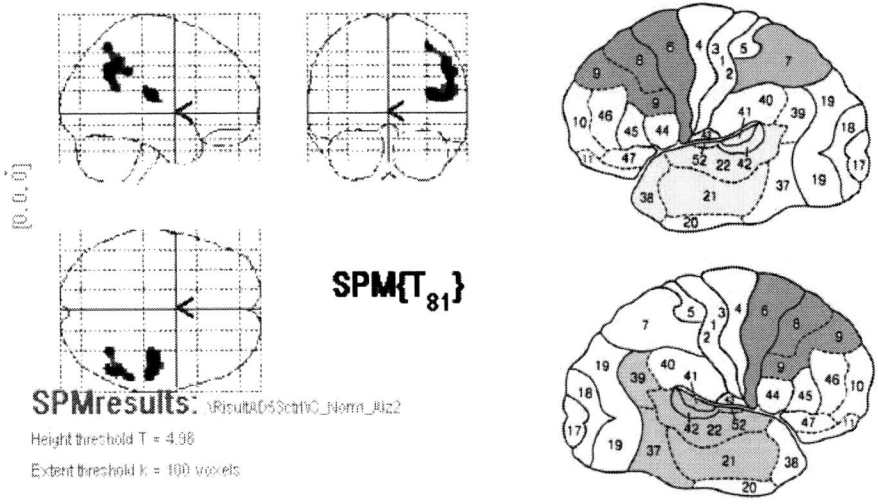

Figure 1. Principal Component Analisis (PCA). Differences in rCBF distribution between a group of early Alzheimer patients (n=30) and a group of healthy controls (n=53). PCA (multivariate analysis, on the right) highlights more and larger regions as compared to SPM (univariate analysis). The data on the right have been imported and processed by the Computerised Brain Atlas. Regions are numbered according to Brodmann Areas (BA). The depicted colors correspond to Principal Components highly correlating different BAs. BA7 is connected to inner regions not shown in the representation of the lateral aspect of the hemispheres.

## Spatial Resolutions and Methodological Dishomogeneities

The reconstructed spatial resolutions in clinical settings are approximately 11 and 7 mm with modern SPECT and PET cameras, respectively. The choice of radiotracer and interpretation method may thus be of more importance than the technique. One of the most sensitive points when comparing nuclear medicine studies as performed by different camera systems is the spatial resolution achieved by each of them. The spatial resolution of the SPECT cameras varies between 6.5 and 20 mm. Studies in which spatial resolutions differ as much as three times are difficult to compare. Some differences that could be present in a group of patients may be hidden in small regions when analyzed with a camera system achieving low resolutions. The same holds true for the choice of the reference region. Irrespectively of the debatable involvement of cerebellum in depression and AD, the choice of a portion of it (typically high flow grey matter), or of the voxels with the maximal count density as reference region, results in an underestimation of both high and low count regions that may possibly hide some differences. In fact, the inclusion of the denominator of the normalization ratio of a high number dumps the differences between the various factors. On the other hand, it has to be taken into account that such conservative choice avoids the inclusion of normalization factors from regions with pathological low tracer uptake, and hence it is advisable in cases in which regions with low tracer uptake are expected.

Other factors that might interfere with an appropriate comparison across different studies include the size of the ROIs, the kind of statistical approach used, and having the analysis of data in 2-D instead of 3-D. The use of smaller ROIs and of the 3-D approach is more likely to catch significant differences. This has particular importance in cases where increases of tracer uptake distribution are typically limited to a small region.

# MAJOR DEPRESSIVE DISORDER

## Characteristics of Depression

Major Depressive Disorder (MDD) is an idiopathic condition (i.e. arising spontaneously, or from an unknown cause) that is defined by international psychiatric classification systems by depressed mood, diminished interest or pleasure in activities, alterations of sleep, fatigue, neurocognitive symptoms, guilt feelings, and/or thoughts of death. The symptoms must be present for at least two weeks and cause significant distress in important areas of functioning. It has been suggested that MDD episodes are the most severe state of illness representing only the tip of the iceberg in a common, chronic and disabling disease with alternating symptom severity [33].

## Prevalence of Depression

The lifetime prevalence of MDD has been reported to be as high as 26% [34]. MDD is now included among the ten leading diseases for global disease burden and has been estimated to become the second most important by the year 2020 [35, 36].

## Possible Causes of Depression: The Serotonin and Kindling Models

For several decades, the prevailing hypothesis on depression has been that it is caused by an absolute or relative deficiency of monoamines, especially serotonin. Research has since failed to confirm any serotonergic lesion in depression. Subtypes of depression have recently been suggested to be caused by a deficiency of dopamine, another monoamine [37]. However, a meta-analysis of monoamine depletion studies investigating the etiological link of monoamines with depression failed to demonstrate a direct causal relationship. It was suggested that the monoamine systems might be important systems in the *vulnerability* to become depressed (authors' italic) [38]. The significance of the effects on serotonin caused by the selective serotonin reuptake inhibitors (SSRIs) in the amelioration of depression is unknown [39].

Recent evidence suggests that SSRIs have anticonvulsant effects [40]. The "kindling hypothesis of mood disorders" was developed in the 1980s on the basis of kindling in epilepsy (i.e. sensitization of brain tissue, particularly the limbic regions, the amygdala and the hippocampus, to electrical current or convulsing drugs resulting in seizure activity when sub-threshold stimuli are applied), stress sensitization research, and the observed inhibitory effects of antiepileptic drugs on episodes of mood symptoms that are initially stress related but eventually appear to occur autonomously [41]. In a study on the effect of stressful life events in MDD, a "pre-kindling model" was found to have the best fit with the data implying that a genetic contribution is associated with the predilection to develop spontaneous depressive episodes [42].

## The Mitochondrial Model of Depression

Mitochondrial energy metabolism (i.e. the production of adenosine triphosphate, ATP) is an extremely complicated process that is affected by the expression of over a thousand proteins, whose genes are encoded on the chromosomes and the maternally-inherited mitochondrial DNA. Mitochondrial dysfunction has been implicated in the pathogenesis of several neurological diseases including AD and PD [43]. Multiple somatic symptoms, and depression, are common in the mitochondrial disorders in which there is a genetic-based deficiency in the production of ATP [44, 45, 46]. Stress-induced precipitation, possibly linked to increasing energy demands, occurs in patients with mitochondrial disorders [47].

Signs of mitochondrial dysfunction have been demonstrated in depression by $^{31}$phosphorous magnetic resonance spectroscopy [48, 49], respiratory chain enzymology and ATP production rates, and by an increased prevalence of small deletions of the mitochondrial DNA [50]. A several-fold increased likelihood of developing depression can be maternally inherited along with the mitochondrial DNA [51, 52], which strongly argues that mitochondrial dysfunction can precede and predispose to depression. A review a studies of mitochondrial alterations in mood disorders has been published recently [53].

As brain has a high ATP requirement in order to maintain the transmembrane potential and neurotransmitter signaling, there are many potential mechanisms as to how energy deficiency might predispose towards mood symptoms. A mouse model with multiple deletions of the mitochondrial DNA demonstrates both mood disorder-like phenotypes and decreased levels and/or turnover of the monoamines serotonin and noradrenalin [54].

Aging, and deficiencies of omega-3 fatty acids and vitamin D, affect mitochondrial function [55, 56, 57]. A protective effect in depression has been suggested for omega-3 and vitamin D [58, 59]. The increasing depression prevalence in the population might reflect several factors including the increase of the elderly, altered nutrient habits, and less out-door activities. Such factors may be of more importance in genetically vulnerable subjects.

## Stress-Induced Depression, and a Case Report

The "impaired adaption" hypothesis of depression proposes an inability of neuronal systems to adequately adapt to aversive stimuli such as stress, and that the effects of antidepressants are exerted through the reconstitution or enhancement of neuronal plasticity [60]. In a study of female subjects on long-term sick leave with depressed moods giving job-related stress as a reason for disability, there was impaired working memory but no differences in hippocampal or prefrontocortical volumes in the comparisons with healthy controls. Prolonged stress exposure, or a pre-existing susceptibility factor, were suggested as possible causes [61].

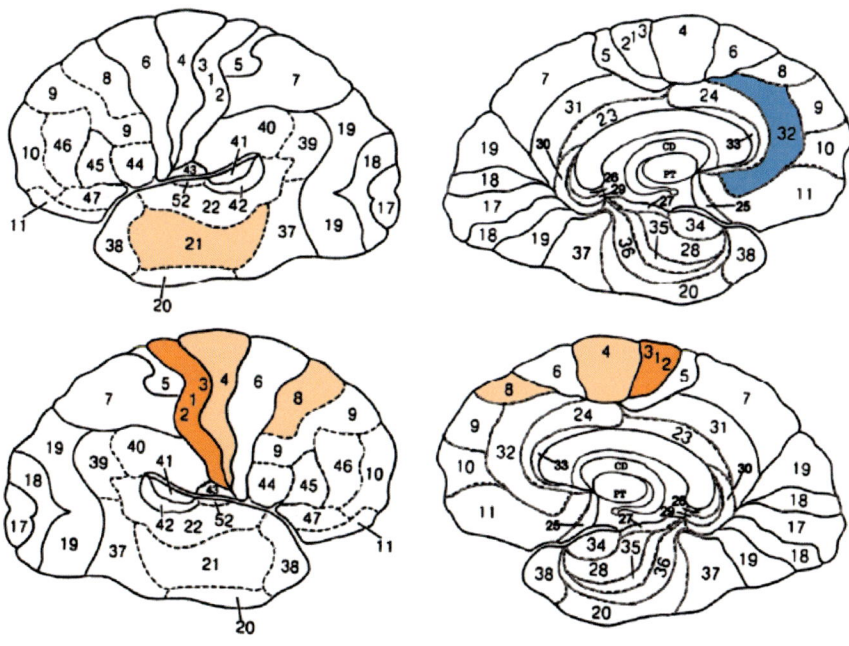

Figure 2. Stress-induced depression. 99mTc-HMPAO SPECT utilizing the CBA in an un-medicated 46-year old female with a five-year-history of chronic depression with pronounced neurocognitive symptoms, tinnitus and muscle pain in comparison with 20 healthy females (mean age 47 ± 9 years). The numeration depicts Brodmann areas (BA). Blue: decreased tracer retention below control mean − 3 standard deviations (SD) in BA 32 in the left anterior cingulate involved in rational thought processes. Orange: increase above + 3 SD in right BAs 1+2+3 in the primary somatosensory cortex. Light orange: increases above + 2 SD in BA 21 in the left temporal lobe participating in auditory processing, and in the right frontal lobe BA 4 in the primary motor cortex, and in BA 8 linked to the experience of uncertainty. No alterations of tracer retention outside of control mean ± 2 SD were observed in another female patient with stress-induced chronic depression without overt neurocognitive symptoms, tinnitus or muscle pain.

In Figure 2, we present the CBA analysis of $^{99m}$Tc-HMPAO SPECT obtained in a female patient five years after the onset of depression, with ensuing chronicity of neurocognitive impairment, muscle pains, and tinnitus and clumsiness at any exertion. Before onset, she was healthy apart from some neck and shoulder stiffness, and had an ambitious but otherwise normal personality. She had been exposed to increasing work-related demands for several years as an administration assistant. In first-degree relatives there was diabetes type II, and decreased pancreatic exocrine function. A pre-existing susceptibility may be present in the patient since muscle stiffness and pain, tinnitus, diabetes, and exocrine pancreatic failure, have been described in mitochondrial disorder [53].

## $^{99m}$Tc-HMPAO SPECT in Depression

In $^{99m}$Tc-HMPAO SPECT studies of un-medicated depression published 1992 – 2001, decreased $^{99m}$Tc-HMPAO retention involving various brain regions, including frontal regions, was reported in several studies [62, 63, 64, 65, 66]. Decreased as well as increased tracer retention [67, 68], or only increased [69], was also reported. No differences with controls were observed in one study [70]. Correlations between $^{99m}$Tc-HMPAO retention in several frontal sub-regions, and depression severity, were found to be negative (lower retention was linked to higher depression rating scores) in one [62], and absent in another study [64].

The results of a metastudy of pooled PET and SPECT studies of depression published before 1999 implicated a general decrease of neuronal activity, and that almost all analyzed cortical and subcortical regions are similarly affected. There were ambiguous results for the limbic system [71].

The results of five $^{99m}$Tc-HMPAO SPECT studies comparing results between depressed patients and healthy controls published since 2002 are presented in Table 1 [72, 73, 74, 75, 76]. Decreased tracer retention in the frontal cortex and other regions was observed in four studies [73, 74, 75, 76]. The increased tracer retention that was observed in one study only may be due to the inclusion of many patients with atypical depression (authors´ personal observation [72]), a subtype of depression characterized by several features including poor response to antidepressant treatment with tricyclic agents.

The results of four studies comparing depressed patients with and without certain clinical features (atypical *versus* non-atypical depression, medication non-responsive *versus* responsive, and depressed patients without *versus* with tinnitus) published since 2002 are presented in Table 2 [72, 74, 76, 77]. There was higher frontal lobe tracer retention in atypical depression [74, 76], in patients non-responsive to a tricyclic antidepressant [77], and in depressed patients without tinnitus [72]. The higher $^{99m}$Tc-HMPAO retention in the patients who were non-responsive to tricyclics [77] may, hypothetically, reflect atypical depression.

It is yet not clear if alterations of tracer retention reflect "trait" phenomena that are present prior to onset of overt depression, or "state" phenomena. Regional increases as well as decreases of tracer retention have been reported at remission in SPECT and PET studies of depression. In a metastudy, the trend for reductions in all cortical and subcortical regions remained for remitted depressed patients [71]. Increased $^{99m}$Tc-HMPAO retention in the left frontal and anterior cingulate gyrus has been reported after electro-convulsive treatment [78], and decreased $^{99m}$Tc-HMPAO retention in the orbitofrontal and/or anterior cingulate after treatment with repetitive transcranial stimulation [79].

**Table 1.** $^{99m}$Tc-HMPAO SPECT studies in patients with depressive disorders and healthy controls

| Study | Year | Depression Type | Size N= | Control Group | Size N= | Design | Regions Involved In Depression | |
|---|---|---|---|---|---|---|---|---|
| | | | | | | | Increase | Decrease |
| Pagani et al [76] | 2007 | Chronic depression with lifetime MDD | 23 | Healthy subjects with a similar gender proportion and mean age | 23 | Resting state SPM | | Right frontal lobe (BAs 6, 8 and 9) |
| Krausz et al [75] | 2007 | MDD before treatment | 10 | Healthy subjects with a similar gender proportion and mean age | 10 | Resting state SPM | | Bilateral left > right in frontal cortex, insula, pre- and post central gyri, superior temporal, inferior parieto-occipital |
| Fountoulakis et al [74] | 2004 | MDD | 50 | Healthy subjects with a similar gender proportion and mean age | 20 | Resting state ROI analyses | | Bilateral temporal and parietal lobes, left occipital lobe, bilateral thalami, left globus pallidus |
| Bonne et al [73] | 2003 | MDD | 23 | Healthy subjects with a similar gender proportion and mean age | 21 | Resting state SPM as well as ROI analyses | | Right parietal and occipital lobes |
| Gardner et al [72] | 2002 | Chronic depression with lifetime MDD | 45 | Healthy subjects with a similar gender proportion and mean age | 26 | Resting state CBA | Right frontal lobe (BAs 8, 9, 10, 45 and 46) | |

INCREASE/DECREASE= increased/decreased $^{99m}$Tc-HMPAO retention in comparison with the control group. BA: Brodmann area.

## Table 2. $^{99m}$Tc-HMPAO SPECT studies between subgroups of depressed patients

| Study | Year | Depression Subtype | Size N= | Depression Subtype | Size N= | Design | Differences Between Subtypes Increase | Differences Between Subtypes Decrease |
|---|---|---|---|---|---|---|---|---|
| Pagani et al [76] | 2007 | Atypical depression | 11 | Non-atypical non-melancholic depression | 12 | Resting state SPM | Atypicals: Bilateral frontal lobe (BAs 6 and 8), parietal lobe (BAs 1-3, 5, 7 and 40) | |
| Fountoulakis et al [74] | 2004 | Atypical depression | 14 | Melancholic depression | 16 | Resting state ROI analyses | Atypicals: Right frontal lobe | Atypicals Right occipital lobe |
| Navarro et al [77] | 2004 | Medication non-responsive MDD | 13 | Medication responsive MDD | 34 | Resting state ROI analyses | Non-responsive: Left anterior frontal lobe | |
| Gardner et al [72] | 2002 | Chronic depression with lifetime MDD without tinnitus | 18 | Chronic depression with lifetime MDD with tinnitus | 27 | Resting state CBA | Non-tinnitus: right frontal lobe (BA 45), left parietal lobe (BA 39), left occipital lobe (BA 18) | |

INCREASE/DECREASE= increased/decreased $^{99m}$Tc-HMPAO retention in comparison to the other subgroup/condition. BA: Brodmann area.

## Origin of Decreased $^{99m}$Tc-HMPAO Retention in Depression

It is unknown if the abnormalities detected at functional imaging are pathophysiologically involved in the evolution of depression, or are an additional expression of the as yet unknown etiological factor(s) of the disease. In AD and PD, decreased $^{99m}$Tc-HMPAO retention has been linked to regional cell loss. The decreases of $^{99m}$Tc-HMPAO retention that have been reported in depression may, hypothetically, reflect the spine loss indicated by the decreases of synaptic "products" [80], the decrease of prefrontal inhibitory local circuit neurons [81], and/or the reduced cortical glial cell numbers that have been reported in depression [82]. Decreased activity in (some) functional circuits is another possibility.

## Origin of Increased $^{99m}$Tc-HMPAO Retention in Depression

In a group of un-treated depressed in-patients with increases of $^{99m}$Tc-HMPAO retention in the superior left frontal lobe and the inferior right temporal lobe, there was a decrease of the alterations in the frontal lobe after successful antidepressant therapy. The findings were interpreted to suggest that the frontal alterations may represent a "state or episode-dependent marker". Anxiety was deemed to be an unlikely cause since the depressed patients were no more anxious than the bipolar patients in the study, in whom frontal $^{99m}$Tc-HMPAO retention was not increased [69].

Increased $^{99m}$Tc-HMPAO retention in depression may reflect increased activity in (some) functional units as well as other phenomena such as increased intracellular GSH levels.

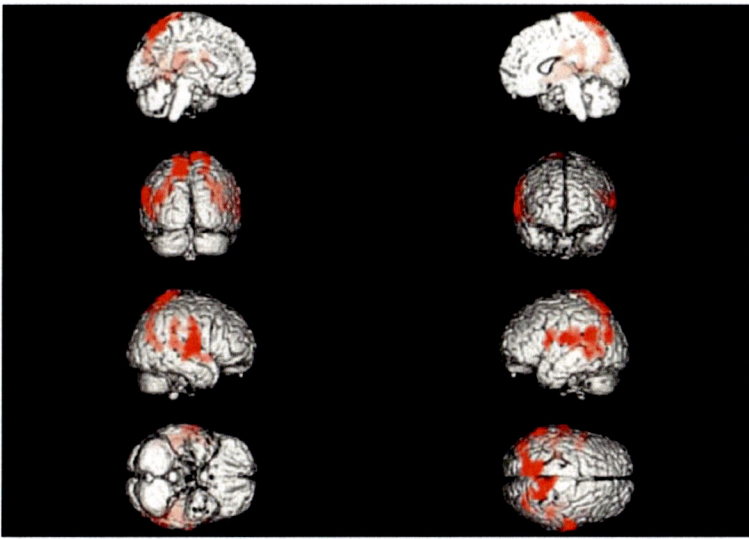

Figure 3. Higher 99mTc-HMPAO retention in depression with mitochondrial dysfunction. Subtraction image depicting the regions in which 99mTc-HMPAO retention in a group of depressed patients showing low mitochondrial cytochrome-oxidase (COX) enzyme activity in muscle (n=11) is higher than in a group of depressed patients with normal COX activity (n=8). The data have been imported and processed by SPM and are depicted as 3D surface rendering.

This phenomenon has been described in the early stages of mitochondrial disease and is considered to be due to increased oxidative stress secondary to reduced respiratory chain enzyme activity [83]. Intracellular metabolic dysfunction is followed by increased rCBF in severe mitochondrial pathology such as Mitochondrial myopathy, Encephalopathy, Lactic Acidosis and Stoke-like episodes (MELAS), thus producing the so-called "luxury perfusion". This phenomenon has been suggested to represent adaptation to the altered mitochondrial function leading to impaired intracellular oxygen utilisation and metabolism. It was speculated that increased CBF was a compensation for increased wash out of lactate produced by increased anaerobic metabolism [84], or might have been related to decreased pH caused by local increase of lactic acid [85]. Quantitative CBF measurements with $^{133}$Xe demonstrated diffuse hyperperfusion in a young man affected by MELAS years prior to undergoing a large posterior "metabolic" stroke [86]. Higher $^{99m}$Tc-HMPAO retention in scattered brain regions has been observed in depressed subjects with decreased muscle activities of mitochondrial enzymes *versus* depressed patients with normal enzyme activities (Gardner and Pagani, submitted manuscript), see Figure 3.

## The Regional Brain Involvement in Depression

Although a frontal lobe involvement has been a common finding $^{99m}$Tc-HMPAO SPECT in depression, alterations of tracer retention are often found, and in some studies only, in other cortical regions and central structures [73, 74, 75].

# POST-TRAUMATIC STRESS DISORDER

## Characteristics of PTSD

Post-traumatic stress disorder (PTSD) is a clinical condition that may affect victims of major psychological trauma and is one of the major contributors of mental suffering [87]. PTSD was defined in DSM-III in 1980 and is a dysfunctional learning with derangement of memory and mood regulation leading to a conditioned fear response elicited by internal or external cues associated with the traumatic situation.

The traumatic event is recalled in flashbacks with involuntary vivid replays, concomitant autonomic reactions and negative feelings. This oppressive tendency to re-experience the trauma leads to avoidance of reminders, irritability and social and emotional withdrawal [88]. The recurring negative trauma memory acts as new trauma-experiences leading to sensitisation of the brain networking engaged in fear response and resulting into the emotional bodily reactions of autonomic arousal.

PTSD is often life-long and is associated with intrusive distressing recollections (flashbacks and/or nightmares), avoidance to stimuli related to trauma and autonomic hyper-reactivity. The development of PTSD is directly correlated to the number of trauma exposures and to the nature of the trauma. Earlier traumatic experiences increase the risk of PTSD development [89] and more severe physical threats result in a higher percentage of PTSD than simply witnessing a traumatic event. The nature of the traumas under investigation in PTSD literature has been mostly represented in the past by combat experience and sexual abuse [90]. However more recent studies have described PTSD in occupational- and society-related stressors [91, 92, 93, 94, 95].

## Prevalence of PTSD

It is estimated that in the general population there is a lifetime prevalence of PTSD of 1.3-9 % [87, 96, 97, 98], making of PTSD the fourth most common psychiatric disorder [98].

## Symptom Provocation and Functional Anatomy

Symptom provocation paradigms are an extremely useful and are a powerful way of delineating the functional anatomy of the traumatic memory that characterizes PTSD. Alterations of local activations at specific tasks indicate dysfunctions of neural processing. In this respect autobiographical trauma-script exposure [92, 93, 99] or audio and visual trauma-related stimuli [100, 101] are a valid approach to elicit rCBF changes in PTSD and the improvement in both technical capabilities and methodology has rendered neuroimaging studies particularly suitable in investigating *in vivo* the neurobiology of emotions. It has to be also taken into consideration that neutral script, being a procedure new to the patient, might also rise the stress levels and/or distract attention as well as resting state that might not result to be completely identical across different investigations. Both these fact are partially responsible for the inconsistent results in PTSD research.

## Functional Imaging in PTSD

Studies performed with both SPECT or PET have reported regional cerebral blood flow changes during trauma recall. Both increases [100, 101, 102, 103, 104, 105, 106] and decreases [100, 102, 103, 105, 106, 107, 108, 109] in rCBF were described. Altered rCBF were mostly found in hippocampus, amygdala, orbito-frontal and frontal cortex, middle prefrontal and middle temporal cortex, anterior and posterior cingulated cortex, temporal poles, Broca's area and caudate. The critical involvement of limbic system has been hypothesized to be connected to the fear-related stimuli and to the emotional responsiveness to the retrieved traumatic experience elicited by symptom provocation.

## $^{99m}$Tc-HMPAO SPECT in PTSD

In this review we took into consideration all English language articles published in the last decade about brain changes in patients with the diagnosis PTSD, mostly assessed by $^{99m}$Tc-HMPAO SPECT (Table 3). Four investigations in which the effect of treatment on PTSD was reported are also discussed (Table 4).

Table 3. $^{99m}$Tc-HMPAO SPECT studies in Post-Traumatic Stress Disorder (PTSD)

| Study | Year | Traumatic Event | Size N= | Control Group | Size N= | Design | Regions Involved In PTSD Increase | Regions Involved In PTSD Decrease |
|---|---|---|---|---|---|---|---|---|
| Chung et al | 2006 | Civil trauma | 23 | Healthy subjects | 64 | Resting state | Cerebellum, Limbic structures, frontal lobe, hippocampus | Parietal lobe, pre-motor cortex, temporal lobe |
| Pagani et al | 2005 | Unintentional killing and assaultive | 20 | Same trauma without PTSD | 27 | Trauma script | Right hemisphere | |
| Lindauer et al | 2004 | Police duties | 15 | Same trauma without PTSD | 15 | Trauma script | Right cuneus | Superior temporal, left middle frontal and left inferior frontal gyri. |
| Bonne et al | 2003 | Motor vehicle accidents | 11 | Civilian trauma without PTSD and healthy subjects | 28 | Resting state | Cerebellum, right precentral, superior and inferior temporal, fusiform, supra-marginal and post-central gyri | |
| Pavic et al | 2003 | Combat-related | 25 | Interhemispheric asymmetry > 20% | 12 | Trauma script | Dominant hemisphere and nc accumbens | |
| Mirzaei et al | 2001 | Torture survivors | 8 | Healthy subjects | 8 | Resting state | Left hemisphere | Temporoparietal cortex |
| Sachinvala et al | 2000 | Mixed traumas | 17 | Healthy subjects | 8 | Resting state | Anterior/posterior cingulate, right temporal and parietal cortex, right basal ganglia, left orbital cortex and hippocampus | |
| Zubieta et al | 1999 | Combat-related | 12 | Same trauma without PTSD and healthy subjects | 23 | Combat sounds | Medial prefrontal cortex | |
| Liberzon et al | 1999 | Combat-related | 14 | Same trauma without PTSD and healthy subjects | 25 | Combat sounds | Left amygdale and nc. accumbens | |

INCREASE/DECREASE= increased/decreased $^{99m}$Tc-HMPAO retention in comparison with the control group.

## Table 4. Changes at SPECT following therapy in Post-Traumatic Stress Disorder (PTSD)

| Study | Year | Traumatic Event | Responders N= | Design | Therapy | Regional Changes Following Therapy Increase | Regional Changes Following Therapy Decrease |
|---|---|---|---|---|---|---|---|
| Pagani et al | 2007 | Unintentional killing and assaultive | 11/15 | Trauma-script | Eye movement desensitization and reprocessing | Lateral temporal lobe (BA 21) | Uncus (BA 36) |
| Peres et al | 2007 | Accident, sexual violence and robbery | NK/16 | Trauma-script | Exposure-based and cognitive restructuring | Parietal lobes (BAs 7 and 40), thalamus, left pre-frontal cortex (BAs 10 and 44), anterior cingulated (BA 32) and hippocampus | Left amygdala |
| Lansing et al | 2005 | Police duties | 6/6 | Trauma-script | Eye movement desensitization and reprocessing | Left frontal gyrus (BA 11, 44, 8 and 9) | Occipital lobe (BA 18), left parietal lobe (BA 40) and right pre-central frontal lobe (BA 4) |
| Seedat et al | 2004 | Civilian and combat-related | 6/11 | Resting state | SRRI (citalopram) | | Left medial temporal cortex (BA 28) |

INCREASE/DECREASE= increased/decreased radiopharmaceutical retention following therapy. NK=Not Known.

It is worth noting that chronic PTSD is often associated with long-term pharmacological treatment and/or alcohol and substance abuse further affecting brain structures and function and in case confounding the results of the investigations. Comparing data reported in Table 3 to a recent extensive review by Francati et al [110] in which 30 PET and fMRI study on PTSD were reviewed, it is worth noting that in general SPECT studies include a larger patients sample (on average 16 vs 9) and have a broader spectrum of traumatic events. In fact of the reviewed $^{99m}$Tc-HMPAO SPECT studies 4/8 did not include combat related or sexual abuse studies whereas this was true only for 5/30 PET and fMRI studies. This is partially due to both the lower spatial and time resolution of SPECT as compared to PET and fMRI leading to the need of a larger number of investigated subjects to reach comparably reliable results. On the other hand, after the large amount of PTSD studies performed in the past on veterans and abused women and children, nowadays there is a tendency to investigate traumas more related to the daily life and to societal problems.

The control groups utilized in the reviewed studies were either subjects suffering the same trauma without developing PTSD [92, 93, 94; Figure 4], healthy subjects [111, 112], or a mixture of both [94, 101, 102]. The choice of the control group is a critical step of the global analysis in neuroimaging. In PTSD subjects that were exposed to the same trauma as patients without suffering any symptom are likely the best ones to be compared to the group under study since the CBF distribution differences found in group comparisons are completely related to the disorder itself and are not confounded by possible group discrepancies nor biased by other variables.

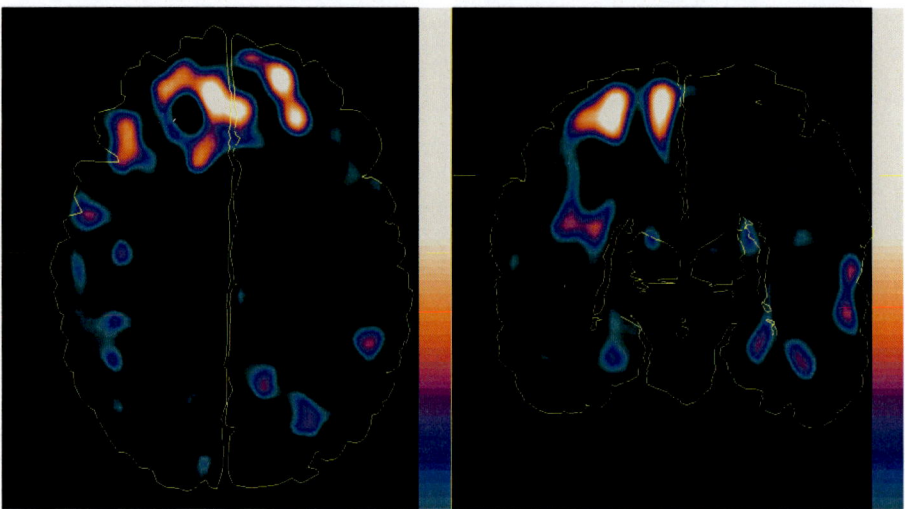

Figure 4. Post Traumatic Stress Disorder. Subtraction image depicting the regions in which 99mTc-HMPAO retention in symptomatic patients showed a higher response to a script-driven stimulus (n=20) than in subjects that underwent the same traumatic experience without developing PTSD (n=27). The data have been imported and processed by the Computerised Brain Atlas and are depicted as transversal (left) and coronal (right) projections.

The most of the $^{99m}$Tc-HMPAO SPECT studies taken into consideration showed a regional increased radiopharmaceutical distribution in PTSD patients as compared to controls. In two cases the increased distribution involved an entire hemisphere [92, 111]. Since $^{99m}$Tc-HMPAO SPECT is a semiquantitative technique it is difficult to definitively rule out the hypothesis that both increased and decreased radiopharmaceutical distribution could represent decreased and increased, respectively, of the $^{99m}$Tc-HMPAO uptake in the rest of the brain. Only three studies reported a regional decreased $^{99m}$Tc-HMPAO distribution, in both cases concomitant to a regional increase in other brain areas [91, 93, 111]. Hemispheric lateralization involved both right and left hemisphere [92, 99, 111] and some of their functions seem to be specifically impaired in PTSD.

The right hemisphere integrates sensory modalities, processes nonverbal emotional communication and seems to be strongly connected with the amygdala, from which receives incoming stimuli of fear and hostility, and the relative regulation of autonomic and hormonal responses. It is also responsible for intrusive emotional memory component of PTSD and autobiographical memories [113]. The left hemisphere generates symbolic representations by categorizing stimuli and personal experiences into novel images and symbols and labels perceptions.

From the analysis of the regions involved in PTSD appears that a large part of the limbic system (hippocampus, fusiform gyrus, amygdala and nc. Accumbens, lentiform nucleus, anterior cingulated and orbitofrontal cortex) plays a key role in the regulation of emotions and storage and retrieval of memoriesThe reported dysfunctions in governing cortices and medial prefrontal cortex are likely related to alterations in planning, execution, inhibition of responses and extinction of fear resulting in motor responses, and in effects on peripheral, sympathetic and cortisol systems (54). Furthermore in PTSD patients changes in prefrontal cortex, essential for encoding and retrieving verbal memories, result in the difficulties in cognitively restructuring traumatic experiences [90].

Considering the key role of this structure in PTSD, it is worth noting the amygdala has been suggested to be mainly activated during recognition and induction of emotions by visual stimuli rather that during the reaction to recalled stimuli [114]. In general in PTSD patients limbic hyperactivation is paralleled by higher cortical hypofunction [115] resulting in a lack of inhibition of reaction to fear from amygdala and lack of adequate attenuation of peripheral sympathetic and hormonal responses to stress. Hyperperfusion and hyperactivity of limbic and paralimbic regions have been suggested to be related to stress-induced long-term potentation between amygdala and periacqueducal grey through N-methyl-D-aspartate (NMDA)-mediated pathway, once a sufficient amount of glutamate is released following stressing events [90]. In PTSD psychological hyperactivation of an overlearned survival response with intrusive phenomena representing an active reworking of trauma memories at cognitive level has also been hypothesized by Silove [116].

Posterior cingulated, parietal and motor cortices are functionally related to prefrontal cortex and mediate cognitive functions of the visuospatial processing, critical in extreme stress situations. Lentiform nucleus, consisting of the putamen and the globus pallidus, was found by Lindauer et al [93] to show a decreased $^{99m}$Tc-HMPAO uptake distribution possibly resulting in reduced motor activities and fewer active coping reactions [117].

The cerebellar involvement in PTSD has been described in two studies [91, 94]. Cerebellum is involved in the autonomic regulation of both cardiovascular and skin responses and emotional behavior as well as in motor conditioning response. Patients with cerebellar dysfunction have a diminished heart rate reduction upon fear conditioning [118] and an abnormal startle response in PTSD [119]. This might result in abnormal delay in habituation, increased heart rate variability, exaggerated startle response and sleep abnormalities [94]. These findings are relevant even methodologically since suggest that the use of cerebellum as reference region for data normalization in PTSD might be inappropriate. The involvement in PTSD of regions processing learning, preparation and execution of motor tasks [94] results in an increased basal level of anxiety and arousal through a continuous preparatory motor activation [94].

Nucleus accumbens, also known as ventral striatum, was found to be activated in two studies [99, 102]. It receives inputs from amygdala, prefrontal cortex and subcubiculum and projects to basal forebrain, diencephalons and rostral brainstem. Its functional-anatomic location makes it involved in creating appropriate emotional responses to incoming stimuli and sensitive to aversive incoming signals. In turn, it modulates the activity in downstream brain structures affecting in PTSD the motor aspect of adaptive responding resulting in exaggerated reactions to incoming information (i.e. hyperirritability, hypersensitivity and hyperexcitability).

## Effect of Therapy on PTSD

Among the four SPECT studies we are aware of in which the effect of therapy on PTSD has been investigated by neuroimaging (see Table 4) two *dealt* with eye movement desensitization and reprocessing [EMDR; 95, 120], one with selective serotonin reuptake inhibitor [SSRI; 121], and one with cognitive restructuring therapy [122].

EMDR is an eclectic therapy method utilizing, among other techniques, relaxation exercises, safe place exercises, cognitive restructuring, future projections and imaging of the trauma combined with positive sensory stimulation. Cognitive restructuring was an exposure-based therapy consisting of an introductory session, an anamnesis session and three restructuring sessions. The SSRI citalopram was administered for 8 weeks at 20 mg/day (first two weeks) and 40 mg/day (last 6 weeks). The therapy in three studies [120, 121, 122] lasted 8 week and it was a little bit longer in the Lansing et al study (on average about 10 weeks).

As reported in Table 4 the traumatic events were very heterogeneous between studies and the design of the study included an individual trauma script during SPECT in three out of four studies [95, 120, 122]. As for the effect of therapies on CBF, both increases and decreases were reported. Probably due to the large variety of traumatic events and therapies the patients underwent to, the regional changes were distributed in almost all cortex (see Table 4). In general in PTSD and in anxiety disorders deactivations are considered to be related to symptom relief and to the normalization of reciprocal associative circuitry regulation and of hyperreactivity in emotional and memory disturbances. Activations were seen as an effect of therapies in improving negative symptoms as depression [95, 120] or as a better inhibition of feed-back processes related to amygdala activity [122].

## CONCLUSION

We, a neurophysiologist implementing nuclear medicine techniques and a clinical psychiatrist, believe that functional neuroimaging may come to play a substantial role in identifying the neurobiological substrates, the biochemical alterations, and the subtypes of psychiatric disorders. The time for an international consensus protocol for functional neuroimaging studies of depression and other psychiatric disorders is approaching.

Multiple somatic symptoms have been suggested to be a core component of the depressive syndrome [123]. The neural correlates of the common somatic symptoms in depression such as pain, tinnitus, and bowel symptoms, may contribute to the "mosaic" of increased and decreased $^{99m}$Tc-HMPAO retention involving various brain regions that has been described in depression [124]. The differences of the $^{99m}$Tc-HMPAO retention reported between depressed patients with and without tinnitus all affected brain regions previously implicated as neural correlates of tinnitus [72]. For this reason, we suggest that depressed patients ought to be divided into subpopulations with shared clinical characteristics in future functional neuroimaging studies.

Applying PCA to MDD patients, increased $^{99m}$Tc-HMPAO retention was found in regions sharing close anatomical and functional relationships [125] confirming the importance of multivariate analysis in detecting functional changes in psychiatric conditions. The deeper investigation level allowed by multivariate analysis might help in clarifying the fine and hidden relationships between regions not detectable with conventional statistical analysis.

## REFERENCES

[1] Dougall NJ, Bruggink S, Ebmeier KP. Systematic review of the diagnostic accuracy of 99mTc-HMPAO-SPECT in dementia. *Am. J. Geriatr. Psychiatry* 2004; 12:554-70.

[2] Poulin P, Zakzanis KK. In vivo neuroanatomy of Alzheimer's disease: evidence from structural and functional brain imaging. *Brain Cogn.* 2002; 49:220-5.

[3] Devous MD Sr. Functional brain imaging in the dementias: role in early detection, differential diagnosis, and longitudinal studies. *Eur. J. Nucl. Med. Mol. Imaging* 2002; 29:1685-96.

[4] Calhoun V, Pearlson GD. Recent developments in brain imaging of schizophrenia: a selective review. *Neuroscience Imaging* 2007: 1:XXXX

[5] Lassen NA, Andersen AR, Friberg L, et al. The retention of [99mTc]-d,l-HM-PAO in the human brain after intracarotid bolus injection: a kinetic analysis. *J. Cereb. Blood Flow Metab.* 1988; 8:S13-22.

[6] Syed GM, Eagger S, O'Brien J, et al. Patterns of regional cerebral blood flow in Alzheimer's disease. *Nucl. Med. Commun.* 1992; 13:656-63.

[7] Salmon E, Sadzot B, Maquet P, et al. Differential diagnosis of Alzheimer's disease with PET. *J. Nucl. Med.* 1994; 35:391-8.

[8] Bonte FJ, Weiner MF, Bigio EH, et al. Brain blood flow in the dementias: SPECT with histopathologic correlation in 54 patients. *Radiology* 1997; 202:793-7.

[9] Jagust W, Thisted R, Devous MD Sr, et al. SPECT perfusion imaging in the diagnosis of Alzheimer's disease: a clinical-pathologic study. *Neurology* 2001; 56:950-6.

[10] Julin P, Lindqvist J, Svensson L, et al. MRI-guided SPECT measurements of medial temporal lobe blood flow in Alzheimer's disease. *J. Nucl. Med.* 1997; 38:914-9.

[11] Ishii K, Sasaki M, Yamaji S, et al. Paradoxical hippocampus perfusion in mild-to-moderate Alzheimer's disease. *J. Nucl. Med.* 1998; 39:293-8.

[12] Rodriguez G, Nobili F, Copello F, et al. 99mTc-HMPAO regional cerebral blood flow and quantitative electroencephalography in Alzheimer's disease: a correlative study. *J. Nucl. Med.* 1999; 40:522-9.

[13] Nobili F, Copello F, Buffoni F, et al. Regional cerebral blood flow and prognostic evaluation in Alzheimer's disease. *Dement Geriatr. Cogn. Disord.* 2001; 12:89-97.

[14] Huang SC, Carson RE, Hoffman EJ, et al. Quantitative measurement of local cerebral blood flow in humans by positron computed tomography and 15O-water. *J. Cereb. Blood Flow Metab.* 1983; 3:141-53.

[15] Yamaguchi T, Kanno I, Uemura K, et al. Reduction in regional cerebral metabolic rate of oxygen during human aging. *Stroke* 1986; 17:1220-8.

[16] Houston AS, Kemp PM, Macleod MA, et al. Use of significance image to determine patterns of cortical blood flow abnormality in pathological and at-risk groups. *J. Nucl. Med.* 1998; 39:425-30.

[17] Imran MB, Kawashima R, Awata S, et al. Parametric mapping of cerebral blood flow deficits in Alzheimer's disease: a SPECT study using HMPAO and image standardization technique. *J. Nucl. Med.* 1999; 40:244-9.

[18] Pagani M. Advances in brain SPECT. *Methodological and human investigations* [thesis]. Departments of Radiology and Hospital Physics, Karolinska Hospital and Karolinska Institutet, Stockholm, Sweden; 2000.

[19] Neirinckx RD, Canning LR, Piper IM, et al. Technetium-99m d,l-HM-PAO: a new radiopharmaceutical for SPECT imaging of regional cerebral blood perfusion. *J. Nucl. Med.* 1987; 28:191-202.

[20] Babich JW. Technetium-99m-HMPAO retention and the role of glutathione: the debate continues. *J. Nucl. Med.* 1991; 32:1681-3.

[21] Ballinger JR, Reid RH, Gulenchyn KY. Technetium-99m HM-PAO stereoisomers: differences in interaction with glutathione. *J. Nucl. Med.* 1988; 29:1998-2000.

[22] Neirinckx RD, Burke JF, Harrison RC, et al. The retention mechanism of technetium-99m-HM-PAO: intracellular reaction with glutathione. *J. Cereb. Blood Flow Metab.* 1988; 8:S4-12

[23] Jacquier-Sarlin MR, Polla BS, Slosman DO. Oxido-reductive state: the major determinant for cellular retention of technetium-99m-HMPAO. *J. Nucl. Med.* 1996; 37:1413-6.

[24] Sasaki T, Fujibayashi Y, Senda M. Distribution of glutathione and technetium-99m-meso-HMPAO in normal and diethyl maleate-treated mouse brain mitochondria. *J. Nucl. Med.* 1998; 39:2178-83.

[25] Fujibayashi Y, Taniuchi H, Waki A, et al. Intracellular metabolism of 99mTc-d,l-HMPAO in vitro: a basic approach for understanding the hyperfixation mechanism in damaged brain. *Nucl. Med. Biol.* 1998; 25:375-8.

[26] Jones K, Johnson KA, Becker JA, et al. Use of singular value decomposition to characterize age and gender differences in SPECT cerebral perfusion. *J. Nucl. Med.* 1998; 39:965-73.

[27] Friston KJ, Frith CD, Liddle PF, et al. The relationship between global and local changes in PET scans. *J. Cereb. Blood Flow Metab.* 1990; 10:458-66.

[28] Strother SC, Anderson JR, Schaper KA, et al. Principal component analysis and the scaled subprofile model compared to intersubject averaging and statistical parametric

mapping: I. "Functional connectivity" of the human motor system studied with [15O]water PET. *J. Cereb. Blood Flow Metab.* 1995; 15:738-53.

[29] Houston AS, Kemp PM, Macleod MA, et al. Use of significance image to determine patterns of cortical blood flow abnormality in pathological and at-risk groups. *J. Nucl. Med.* 1998; 39:425-30.

[30] Friston KJ, Holmes AP, Worsley KJ, et al. Statistical parametric maps in functional imaging: A general linear approach. *Hum. Brain Mapp.* 1995; 2:189-210.

[31] Egan JP. Signal detection theory and ROC analysis. Academic Press, New York 1975.

[32] Swets JA. Indices of discrimination or diagnostic accuracy: their ROCs and implied models. *Psychol Bull* 1986; 99:100-17.

[33] Judd LL, Akiskal HS, Maser JD, et al. A prospective 12-year study of subsyndromal and syndromal depressive symptoms in unipolar major depressive disorders. *Arch. Gen. Psychiatry* 1998; 55:694-700.

[34] Levitt AJ, Boyle MH, Joffe RT, et al. Estimated prevalence of the seasonal subtype of major depression in a Canadian community sample. *Can. J. Psychiatry* 2000; 45:650-4.

[35] Murray CJ, Lopez AD. Alternative projections of mortality and disability by cause 1990-2020: Global Burden of Disease Study. *Lancet* 1997; 349:1498-504.

[36] Lopez AD, Mathers CD, Ezzati M, Jamison DT, Murray CJ. Global and regional burden of disease and risk factors, 2001: systematic analysis of population health data. *Lancet* 2006; 367:1747-57.

[37] Dunlop BW, Nemeroff CB. The role of dopamine in the pathophysiology of depression. *Arch. Gen. Psychiatry* 2007; 64:327-37.

[38] Ruhé HG, Mason NS, Schene AH. Mood is indirectly related to serotonin, norepinephrine and dopamine levels in humans: a meta-analysis of monoamine depletion studies. *Mol. Psychiatry* 2007; 12:331-59.

[39] Lacasse JR, Leo J. Serotonin and depression: a disconnect between the advertisements and the scientific literature. *PLoS Med.* 2005; 2:e392.

[40] Richman A, Heinrichs SC. Seizure prophylaxis in an animal model of epilepsy by dietary fluoxetine supplementation. *Epilepsy Res.* 2007; 74:19-27.

[41] Monroe SM, Harkness KL. Life stress, the "kindling" hypothesis, and the recurrence of depression: considerations from a life stress perspective. *Psychol Rev.* 2005; 112:417-45.

[42] Kendler KS, Thornton LM, Gardner CO. Genetic risk, number of previous depressive episodes, and stressful life events in predicting onset of major depression. *Am. J. Psychiatry* 2001; 158:582-6.

[43] Roses AD, Saunders AM, Huang Y, et al. Complex disease-associated pharmacogenetics: drug efficacy, drug safety, and confirmation of a pathogenetic hypothesis (Alzheimer's disease). *Pharmacogenomics J* 2007; 7:10-28.

[44] Fadic R, Johns DR. Clinical spectrum of mitochondrial diseases. *Semin. Neurol.* 1996; 16:11-20.

[45] Chinnery PF, Turnbull DM. Mitochondrial medicine. *Quart. J. Med.* 1997; 90:657-67.

[46] Fattal O, Link J, Quinn K, et al. Psychiatric comorbidity in 36 adults with mitochondrial cytopathies. *CNS Spectr.* 2007; 12: In press.

[47] Wong LJ, Boles RG. Mitochondrial DNA analysis in clinical laboratory diagnostics. *Clin. Chim. Acta* 2005; 354:1-20.

[48] Moore CM, Christensen JD, Lafer B, et al. Lower levels of nucleoside triphosphate in the basal ganglia of depressed subjects: a phosphorous-31 magnetic resonance spectroscopy study. *Am. J. Psychiatry* 1997; 154:116-8.

[49] Volz HP, Rzanny R, Riehemann S, et al. 31P magnetic resonance spectroscopy in the frontal lobe of major depressed patients. *Eur. Arch. Psychiatry Clin. Neurosci.* 1998; 248:289-95.

[50] Gardner A, Johansson A, Wibom R, et al. Alterations of mitochondrial function and correlations with personality traits in selected major depressive disorder patients. *J. Affect Disord.* 2003; 76:55-68.

[51] Boles RG, Burnett BB, Gleditsch K, et al. A high predisposition to depression and anxiety in mothers and other matrilineal relatives of children with presumed maternally inherited mitochondrial disorders. *Am. J. Med. Genet. B Neuropsychiatr. Genet* 2005; 137:20-4.

[52] Burnett BB, Gardner A, Boles RG. Mitochondrial inheritance in depression, dysmotility and migraine? *J. Affect. Disord.* 2005; 88:109-16.

[53] Gardner A, Boles RG. Is a "mitochondrial psychiatry" in the future? *A review. Curr. Psychiat. Reviews* 2005; 1:255-71.

[54] Kasahara T, Kubota M, Miyauchi T, et al. Mice with neuron-specific accumulation of mitochondrial DNA mutations show mood disorder-like phenotypes. *Mol. Psychiatry* 2006; 11:577-93, 523

[55] Short KR, Bigelow ML, Kahl J, et al. Decline in skeletal muscle mitochondrial function with aging in humans. *Proc. Natl. Acad. Sci. USA* 2005; 102:5618-23.

[56] Flachs P, Horakova O, Brauner P, et al. Polyunsaturated fatty acids of marine origin upregulate mitochondrial biogenesis and induce beta-oxidation in white fat. *Diabetologia* 2005; 48:2365-75.

[57] Vaynman S, Ying Z, Wu A, et al. Coupling energy metabolism with a mechanism to support brain-derived neurotrophic factor-mediated synaptic plasticity. *Neuroscience* 2006; 139:1221-34.

[58] Freeman MP, Hibbeln JR, Wisner KL, et al. Omega-3 fatty acids: evidence basis for treatment and future research in psychiatry. *J. Clin. Psychiatry* 2006; 67:1954-67.

[59] Jorde R, Waterloo K, Saleh F, et al. Neuropsychological function in relation to serum parathyroid hormone and serum 25-hydroxyvitamin D levels. The Tromso study. *J. Neurol.* 2006; 253:464-70.

[60] Laifenfeld D, Karry R, Klein E, et al. Alterations in cell adhesion molecule L1 and functionally related genes in major depression: a postmortem study. *Biol. Psychiatry* 2005; 57:716-25.

[61] Rydmark I, Wahlberg K, Ghatan PH, et al. Neuroendocrine, cognitive and structural imaging characteristics of women on longterm sickleave with job stress-induced depression. *Biol. Psychiatry* 2006; 60:867-73.

[62] Yazici KM, Kapucu O, Erbas B, et al. Assessment of changes in regional cerebral blood flow in patients with major depression using the 99mTc-HMPAO single photon emission tomography method. *Eur. J. Nucl. Med.* 1992; 19:1038-43.

[63] Lesser IM, Mena I, Boone KB, et al. Reduction of cerebral blood flow in older depressed patients. *Arch. Gen. Psychiatry* 1994; 51:677-86.

[64] Fischler B, D'Haenen H, Cluydts R, et al. Comparison of 99m Tc HMPAO SPECT scan between chronic fatigue syndrome, major depression and healthy controls: an

exploratory study of clinical correlates of regional cerebral blood flow. *Neuropsychobiology* 1996; 34:175-83.

[65] Tutus A, Kibar M, Sofuoglu S, et al. A technetium-99m hexamethylpropylene amine oxime brain single-photon emission tomography study in adolescent patients with major depressive disorder. *Eur. J. Nucl. Med.* 1998; 25:601-6.

[66] Navarro V, Gasto C, Lomena F, et al. Frontal cerebral perfusion dysfunction in elderly late-onset major depression assessed by 99MTC-HMPAO SPECT. *Neuroimage* 2001; 14:202-5.

[67] Kowatch RA, Devous Sr MD, Harvey DC, et al. A SPECT HMPAO study of regional cerebral blood flow in depressed adolescents and normal controls. *Prog. Neuropsychopharmacol. Biol. Psychiatry* 1999; 23:643-56.

[68] Milo TJ, Kaufman GE, Barnes WE, et al. Changes in regional cerebral blood flow after electroconvulsive therapy for depression. *J. ECT* 2001; 17:15-21.

[69] Tutus A, Simsek A, Sofuoglu S, et al. Changes in regional cerebral blood flow demonstrated by single photon emission computed tomography in depressive disorders: comparison of unipolar vs. bipolar subtypes. *Psychiatry Res.* 1998; 83:169-77.

[70] Maes M, Dierckx R, Meltzer HY, et al. Regional cerebral blood flow in unipolar depression measured with Tc-99m-HMPAO single photon emission computed tomography: negative findings. *Psychiatry Res.* 1993; 50:77-88.

[71] Nikolaus S, Larisch R, Beu M, et al. Diffuse cortical reduction of neuronal activity in unipolar major depression: a retrospective analysis of 337 patients and 321 controls. *Nucl. Med. Commun.* 2000; 21:1119-25.

[72] Gardner A, Pagani M, Jacobsson H, et al. Differences in resting state regional cerebral blood flow assessed with 99mTc-HMPAO SPECT and brain atlas matching between depressed patients with and without tinnitus. *Nucl. Med. Commun.* 2002; 23:429-39.

[73] Bonne O, Louzoun Y, Aharon I, et al. Cerebral blood flow in depressed patients: a methodological comparison of statistical parametric mapping and region of interest analyses. *Psychiatry Res.* 2003; 122:49-57.

[74] Fountoulakis KN, Iacovides A, Gerasimou G, et al. The relationship of regional cerebral blood flow with subtypes of major depression. *Prog. Neuropsychopharmacol. Biol. Psychiatry* 2004; 28:537-46.

[75] Krausz Y, Freedman N, Lester H, et al. Brain SPECT study of common ground between hypothyroidism and depression. *Int J Neuropsychopharmacol* 2007;10:99-106.

[76] Pagani M, Salmaso D, Nardo D, et al. Imaging the neurobiological substrate of atypical depression by SPECT. *Eur. J. Nucl. Med. Mol. Imaging* 2007; 34:110-20.

[77] Navarro V, Gasto C, Lomena F, et al. Prognostic value of frontal functional neuroimaging in late-onset severe major depression. *Br. J. Psychiatry* 2004; 184:306-11.

[78] Vangu MD, Esser JD, Boyd IH, et al. Effects of electroconvulsive therapy on regional cerebral blood flow measured by 99mtechnetium HMPAO SPECT. *Prog. Neuropsychopharmacol. Biol. Psychiatry* 2003; 27:15-9.

[79] Nadeau SE, McCoy KJ, Crucian GP, et al. Cerebral blood flow changes in depressed patients after treatment with repetitive transcranial magnetic stimulation: evidence of individual variability. *Neuropsychiatry Neuropsychol. Behav. Neurol.* 2002; 15:159-75.

[80] Eastwood SL, Harrison PJ. Synaptic pathology in the anterior cingulate cortex in schizophrenia and mood disorders. A review and a Western blot study of synaptophysin, GAP-43 and the complexins. *Brain Res. Bull.* 2001; 55:569-78.

[81] Rajkowska G. Postmortem studies in mood disorders indicate altered numbers of neurons and glial cells. *Biol. Psychiatry* 2000; 48:766-77.

[82] Cotter DR, Pariante CM, Everall IP. Glial cell abnormalities in major psychiatric disorders: the evidence and implications. *Brain Res. Bull.* 2001; 55:585-95.

[83] Filosto M, Tonin P, Vattemi G, et al. Antioxidant agents have a different expression pattern in muscle fibers of patients with mitochondrial diseases. *Acta Neuropathol. (Berl)* 2002; 103:215-20.

[84] Nariai T, Ohno K, Ohta Y, et al. Discordance between cerebral oxygen and glucose metabolism, and hemodynamics in a mitochondrial encephalomyopathy, lactic acidosis, and strokelike episode patient. *J. Neuroimaging* 2001; 11:325-9.

[85] Peng NJ, Liu RS, Li JY, et al. Increased cerebral blood flow in MELAS shown by Tc-99m HMPAO brain SPECT. *Neuroradiology* 2000; 42:26-9.

[86] Rodriguez G, Nobili F, Tanganelli P, et al. Cerebral hyperperfusion antedates by years strokelike episodes in the MELAS syndrome. *Stroke* 1996; 27:341-2.

[87] Kessler, R.C. Posttraumatic stress disorder: the burden to the individual and to society. *J. Clin. Psychiatry* 2000; 61:4-12.

[88] American Psychiatric Association, 1994. *Diagnostic and Statistical Manual of Mental disorder,* ed 4 (DSM-IV). APA, Washington, DC.

[89] Mollica RF, McInnes K, Poole C, et al. Dose-effect relationships of trauma to symptoms of depression and post-traumatic stress disorder among Cambodian survivors of mass violence. *Br. J. Psychiatry* 1998; 173:482-88.

[90] Hull AM. Neuroimaging findings in post-traumatic stress disorder. *Br. J. Psychiatry* 2002; 181:102-10.

[91] Chung YA, Kim SH, Chung SK, et al. Alterations in cerebral perfusion in posttraumatic stress disorder patients without re-exposure to accident-related stimuli. *Clin. Neurophysiol.* 2006; 117:637-42.

[92] Pagani M, Högberg G, Salmaso D, et al. Regional cerebral blood flow during auditory recall in 47 subjects exposed to assaultive and non-assaultive trauma and developing or not Post Traumatic Stress Disorder. *Eur. Arch. Psychiatry Clin. Neurosci.* 2005; 255: 359-65.

[93] Lindauer RJ, Booij J, Habraken JB, et al. Cerebral blood flow changes during script-driven imagery in police officers with posttraumatic stress disorder. *Biol. Psychiatry* 2004; 56:853-61.

[94] Bonne O, Gilboa A, Louzoun Y, et al. Resting regional cerebral perfusion in recent posttraumatic stress disorder. *Biol. Psychiatry* 2003; 54:1077-86.

[95] Lansing K, Amen DG, Hanks C, et al. High-resolution brain SPECT imaging and eye movement desensitization and reprocessing in police officers with PTSD. *J. Neuropsychiatry Clin. Neurosci.* 2005 Fall;17:526-32.

[96] Breslau N. The epidemiology of posttraumatic stress disorder: what is the extent of the problem? *J Clin Psychiatry* 2001; 62: 16-22.

[97] Davidson JRT, Tharwani HM, Connor , et al. Depression Trauma Scale (DTS): normative scores in the general population and effect sizes in placebo-controlled SSRI trials. *Depress. Anxiety* 2002; 15: 75-8.

[98] Breslau N, Davis GC, Andreski P, et al. Traumatic events and posttraumatic stress disorder in an urban population of young adults. *Arch. Gen. Psychiatry* 1991;48:216-22.

[99] Pavic L, Gregurek R, Petrovic R, et al. Alterations in brain activation in posttraumatic stress disorder patients with severe hyperarousal symptoms and impulsive aggressiveness. *Eur. Arch. Psychiatry Clin. Neurosci.* 2003; 253:80-3.

[100] Liberzon I, Taylor SF, Amdur R, et al. Brain Activation in PTSD in response to Trauma-Related Stimuli. *Biol. Psychiatry* 1999; 45: 817–26.

[101] Zubieta JK, Chinitz JA, Lombardi U, et al. Medial frontal cortex involvement in PTSD symptoms:A SPECT study. *J. Psychiatr. Res.* 1999; 33:259–64.

[102] Bremner JD, Narayan M, Staib LH, et al. Neural correlates of memories of childhood sexual abuse in women with and without posttraumatic stress disorder. *Am. J. Psychiatry* 1999; 156: 1787-95.

[103] Pissiota A, Frans O, Fernandez M, et al. Neurofunctional correlates of posttraumatic stress disorder: a PET symptom provocation study. *Eur. Arch. Psychiatry Clin. Neurosci.* 2002; 252:68-75.

[104] Rauch SL, Savage CR, Alpert NM, et al. The functional neuroanatomy of anxiety: a study of three disorders using positron emission tomography and symptom provocation. *Biological. Psychiatry* 1997; 42: 446-52.

[105] Shin LM, Kosslyn SM, McNally RJ, et al. Visual Imagery and Perception in Posttraumatic Stress Disorder. A Positron emission Tomographic Investigation. *Archives of General Psychiatry* 1997; 54: 233-41.

[106] Shin LM, McNally RJ, Kosslyn SM, et al. Regional cerebral blood flow during script-driven imagery in childhood sexual abuse-related PTSD: A PET investigation. *American Journal of Psychiatry* 1999;156: 575-84.

[107] Bremner JD, Staib LH, Kaloupek D. Neural Correlates of Exposure to Traumatic Pictures and Sound in Vietnam Combat Veterans with and without Posttraumatic Stress disorder: A Positron Emission Tomography Study. *Biological Psychiatry* 1999; 45: 806–16.

[108] Bremner JD. Neuroimaging Studies in Post-traumatic Stress Disorder. *Current Psychiatry Reports* 2002; 4: 254–63.

[109] Rauch SL, van der Kolk BA, Fisler RE, et al. A symptom provocation study using Positron emission tomography and Script Driven Imagery. *Archives of General Psychiatry* 1996; 53: 380–87.

[110] Francati V, Vermetten E, Bremner JD. Functional neuroimaging studies in posttraumatic stress disorder: review of current methods and findings. *Depress Anxiety* 2006;24:202-18.

[111] Mirzaei S, Knoll P, Keck A, et al. Regional cerebral blood flow in patients suffering from post-traumatic stress disorder. *Neuropsychobiology* 2001;43:260-4.

[112] Sachinvala N, Kling A, Suffin S, et al. Increased regional cerebral perfusion by 99mTc hexamethyl propylene amine oxime single photon emission computed tomography in post-traumatic stress disorder. *Mil. Med.* 2000;165:473-9.

[113] Freeman TW, Roca V. Gun use, attitudes toward violence, and aggression among combat veterans with chronic posttraumatic stress disorder. *J. Nerv. Ment. Dis.* 2001;189:317-20.

[114] Damasio AR, Grabowski TJ, Bechara A, et al. Subcortical and cortical brain activity during the feeling of self-generated emotions. *Nature Neurosci.* 2000; 3:1049–56.

[115] Garcia R, Vouimba RM, Baudry M, Thompson RF. The amygdala modulates prefrontal cortex activity relative to conditioned fear. *Nature* 1999; 402(6759):294-6.

[116] Silove D. Is posttraumatic stress disorder an overlearned survival response? An evolutionary-learning hypothesis. *Psychiatry* 1998; 61:181-90.

[117] LeDoux JE, Gorman JM. A call to action: overcoming anxiety through active coping. *Am. J. Psychiatry* 2001; 158:1953-5.

[118] Maschke M, Schugens M, Kindsvater K, et al. Fear conditioned changes of heart rate in patients with medial cerebellar lesions. *J. Neurol. Neurosurg Psychiatry* 2002; 72:116-8.

[119] Cohen H, Benjamin J, Geva AB, et al. Autonomic dysregulation in panic disorder and in post-traumatic stress disorder: application of power spectrum analysis of heart rate variability at rest and in response to recollection of trauma or panic attacks. *Psychiatry Res.* 2000; 96:1-13.

[120] Pagani M, Högberg G, Salmaso D, et al. Effects of EMDR psychotherapy on 99mTc-HMPAO distribution in occupation-related Post-Traumatic Stress Disorder. *Nuc. Med. Commun.* 2007; In press.

[121] Seedat S, Warwick J, van Heerden B, Hugo C, Zungu-Dirwayi N, Van Kradenburg J, Stein DJ, et al. Single photon emission computed tomography in posttraumatic stress disorder before and after treatment with a selective serotonin reuptake inhibitor. *J. Affect. Disord.* 2004; 80:45-53.

[122] Peres JF, Newberg AB, Mercante JP, et al. Cerebral blood flow changes during retrieval of traumatic memories before and after psychotherapy: a SPECT study. *Psychol. Med.* 2007; doi:10.1017/S003329170700997X.

[123] Simon GE, VonKorff M, Piccinelli M, et al. An international study of the relation between somatic symptoms and depression. *N. Engl. J. Med.* 1999; 341:1329-35.

[124] Vasile RG, Schwartz RB, Garada B, et al. Focal cerebral perfusion defects demonstrated by 99mTc-hexamethylpropyleneamine oxime SPECT in elderly depressed patients. *Psychiatry Res.* 1996; 7:59-70.

[125] Pagani M, Gardner A, Salmaso D, et al. Principal component and volume of interest analyses in depressed patients imaged by 99mTc-HMPAO SPET: a methodological comparison. *Eur. J. Nucl. Med. Mol. Imaging* 2004; 31:995-100.

*Chapter 10*

# COMPARING 1.5T AND 3T BOLD fMRI IMAGING OF FINGER TAPPING WITH FAMILIAR AND NOVEL SEQUENCES

*Lars Nyberg*[*,1,2, CA,] *Anne Larsson*[1], *Johan Eriksson*[3], *Richard Birgander*[1], *Torbjörn Sundström*[1] *and Katrin Riklund Åhlström*[1]

[1] Department of Radiation Sciences (Diagnostic Radiology)
[2] Department of Integrative Medical Biology (Physiology)
[3] Department of Psychology

## ABSTRACT

It has been suggested that fMRI at 3T yields stronger and more extensive BOLD activations than fMRI at 1.5T, and that imaging at higher field strengths can reveal unique activations. In the present study we compared, within-subjects, activation patterns during a finger-tapping task at 1.5 and 3T. The data were analyzed with a random-effects model in SPM2. At a strict statistical level ($p<0.05$, FWE correction for multiple comparisons), ipsilateral cerebellar activation was revealed at 1.5T. At 3T, activation in sensory-motor regions in the contra-lateral cerebrum was identified in addition to the activation in cerebellum. At a less stringent statistical threshold, imaging at 1.5T and 3T revealed overlapping cortical regions with more extensive clusters at 3T. A similar pattern was seen in a comparison of familiar and novel sequences. However, subcortical activations of thalamus and parts of the basal ganglia were uniquely identified at 3T. Analyses at the individual level substantiated the group results by showing that the higher sensitivity of the 3T resulted in images with higher between-individual consistency in activation patterns.

---

[*] Umeå University, S-901 87 Umeå, Sweden; Lars.Nyberg@diagrad.umu.se

# INTRODUCTION

Functional MRI on 3T scanners is becoming increasingly more common (Voss et al., 2006). Although the number of direct comparisons is still limited, the available data show that fMRI on 3T scanners yields stronger and more extensive activations than 1.5T imaging (Fera et al., 2004; Krüger et al., 2001; Nakai et al., 2001). In a study of manual motor decisions, Hoenig et al. (2005) found that the mean *t* value associated with peak voxels was 1.3 times larger at 3T than at 1.5T. They also found that the number of activated voxels was 60-80% higher at 3T. Krasnow et al. (2003) also found that imaging on 3T produced more extensive activations. Notably, the magnitude of the difference between 1.5 and 3T imaging varied across functional tasks and associated brain regions. In a visual perception task, a 23% increase in number of activated voxels in striate and extrastriatal regions was observed. In a working memory task, the corresponding increase in fronto-parietal voxel activity ranged between 36-83%.

In addition to resulting in stronger and more extensive activations, imaging at 3T seems to reveal activations that are not detected at 1.5T. Such activations were located in prefrontal cortex during the working memory task in the Krasnow et al. (2003) study, and in lateral and medial premotor areas in the Hoenig et al. (2005) study. Between-study comparisons underscore the impression that imaging at higher field strengths can reveal unique activations (see Ugurbil et al., 1999; Voss et al., 2006). Enhanced sensitivity at the individual level for 3T can have clinical implications by reducing false-negative activation changes in individuals that undergo neurosurgical planning (Vlieger et al., 2004).

The aim of the present study was to evaluate, within subjects, differences in brain activation for a pre-trained and a novel finger tapping sequence with 1.5 and 3T fMRI. This manipulation allowed comparison of results acquired at different field strengths for both a "loose" (finger tapping vs. rest) and a "tight" (trained vs. untrained finger tapping) contrast. In addition to group (random-effects) analyses, we evaluated differences at the individual level.

# MATERIALS AND METHODS

## Subjects

Twelve normal, healthy subjects participated in this study (21-34 years of age, 6 women). All subjects were right handed by self report, and had normal or corrected to normal vision. Participants gave written informed consent, and the study was approved by the ethics committee at the University Hospital of Northern Sweden. The subjects were paid for participation.

## Functional Activation Task

The task was modelled after that used in Nyberg et al. (2006). Just before the first scanning session, the subjects were introduced to the task. They were instructed that the left index to little fingers were numbered from 1 to 4 and that they were to perform finger tapping sequences as fast and accurate as possible according to visually presented 5-digit sequences. They practiced for 4×1.5 minutes, with a 1 minute resting period between each training session, on one of two sequences (assignment of sequences as trained or untrained was counterbalanced across subjects). Just before the second scanning session, the subjects had a shorter training period where they were instructed to perform ten finger tapping sequences at their own pace.

In each of the two scanning sessions, trained and untrained sequences were presented visually in a random order according to an event-related design with an inter-stimulus interval of 6.0 s. For each presented sequence, the subjects were to perform the corresponding series of finger tapping as fast and accurate as possible. A total of 40 sequences were presented, half of which were trained and the other half untrained. Twenty resting periods of 6.0 s were randomly distributed between the sequences, during which the subjects viewed "x x x x x" on the screen.

The number of correct sequences performed by each participant were counted, and a correct sequence was defined as a complete motor replicate of the visually presented sequence (e.g., 3 2 4 1 3 or 3 1 4 2 3). Response time was also recorded for each sequence.

## MRI Methods

All subjects performed the finger tapping task on both a 1.5 and a 3T scanner. The scan order was counterbalanced across subjects such that half started with the 1.5T scanner and the other half started with the 3T scanner (for each subject, the two scanning sessions were performed within a period of 4 days). The MRI parameters were kept the same as in our standard fMRI protocols for each scanner (i.e. not specifically adjusted for the comparison between 1.5 and 3T).

*1.5T*: Scanning was conducted on a Philips Intera 1.5 T system (Philips Medical Systems, Netherlands), equipped for echo-planar imaging (EPI). A standard quadrature headcoil was used. Blood-oxygen level-dependent contrast images were acquired using a $T2^*$-weighted single-shot gradient echo EPI sequence with the following parameters: echo time: 50 ms, repetition time: 3000 ms, flip angle: 90°, field of view: 22×22 cm, matrix size: 64×64 and slice thickness: 4.4 mm. Thirty-three transaxial slices positioned to include the whole brain volume were acquired every 3.0 s. Five "dummy scans" were run before the image acquisition started to avoid signals resulting from progressive saturation.

*3T*: These acquisitions were performed on a 3 T Philips Achieva system (Philips Medical Systems, Netherlands), equipped with an 8 channel SENSE head coil. The EPI sequence had the same field of view and matrix size as described above. The other parameters were: echo time: 30 ms, repetition time: 2012 ms, flip angle: 90° and slice thickness 3.4 mm (43 slices acquired every 2.0 s). A SENSE factor of 3.2 was used; a value which previously has been optimized for a similar motor activation task. In this case seven "dummy scans" were run before the start of acquisition.

For both scanners, the finger tapping sequences were presented visually on a semi-transparent screen at the end of the scanner top, which the subject could view via a mirror mounted on the head coil. Cushions inside the head coil were used to reduce head movement, and headphones were used to dampen the scanner noise. The finger tapping task was executed with a four-button response pad on the 1.5T scanner and a five-button response pad on the 3T scanner (only four buttons were used) (Lumitouch reply system, Lightwave Medical Industries, Canada). The response pads were connected to a computer running the program E-prime (Psychology Software Tools, PA, USA) which registered the responses.

## Data Analysis

The fMRI images were transferred to a PC and converted to Analyze format using the program *MRIcro* (Rorden and Brett, 2000). The data was then pre-processed and analysed using in-house developed software (DataZ) and SPM2 (*Wellcome Department of Cognitive Neurology, London, UK*) implemented in Matlab 7.1 (Mathworks Inc., MA, USA). The pre-processing steps were: slice timing correction, realignment with respect to the first image volume in the series, unwarping to reduce residual movement related variance, normalisation to an EPI template in the Montreal Neurological Institute (MNI) space, and smoothing with an isotropic 8.0mm Gaussian filter kernel. Single-subject statistical contrasts were then set up using the general linear model. The sequences of finger tapping were modelled as fixed response (box-car) waveforms convolved with the hemodynamic response function (HRF). The reaction time plus the time required to complete the sequence was averaged for all scanning sessions and used as the duration of the box-car waveforms. Statistical parametric maps (SPMs) were generated using $t$ statistics to identify regions activated according to the model for individual subjects, and random effects analyses were then used to reveal results for the whole group.

We first examined brain activity during finger tapping for both sequences (trained and untrained). In a first step we thresholded the image resulting from this contrast at $p<0.05$ (peak FWE corrected for multiple comparisons). Second, a more lenient threshold of $p<0.001$ (uncorrected for multiple comparisons) was used in order to assess whether activations not seen for 1.5T at the more stringent level would become apparent at the more liberal level. The next contrast compared trained versus untrained finger tapping. We used the same step-wise thresholding approach as in the overall finger-tapping vs. rest contrast. Visualization of results for individual subjects was based on peak voxels from the finger-tapping vs. rest contrast.

To quantify differences in sensitivity between 1.5 and 3T imaging, we used the number of activated voxels as a measure of functional sensitivity (cf., Hoenig et al., 2005). Thus, for commonly activated regions at 1.5T and 3T, we report the number of activated voxels at a given $t$ value.

## RESULTS

### Behavioral Performance

There were no significant differences in accuracy or reaction times for the two scanners. The overall accuracy was high (about 90%) and no significant differences were seen when comparing the performance in the different scanners or for trained and untrained sequences.

The average time required to tap the whole sequences (plus reaction time) was 1.93 s (±0.43) for trained sequences and 2.16 s (±0.48) for untrained sequences in the 1.5T scanner. For the 3T scanner the corresponding values were 2.10 s (±0.62) and 2.32 s (±0.61). Notably, the average difference between the untrained and trained sequences was almost identical on the two scanners (about 0.22 s). This reduction in tapping time for trained sequences constituted evidence of learning, and validated the tight contrast of trained versus untrained tapping.

## Brain Activity during Finger Tapping

Finger tapping on trained and untrained sequences, relative to rest, was only found to activate ipsilateral cerebellum on the 1.5T scanner (Figure 1a). This region was also identified at 3T imaging (Figure 1b), and in addition increased activation was observed in contralateral sensory-motor cortex and SMA (Figure 1c).

When the more lenient statistical threshold was adopted for 1.5T, increased activity was identified in the cortical cerebral regions that were revealed at 3T (Figure 2a). In addition, several areas of the left cerebrum and right cerebellum were revealed at 1.5T at this more liberal threshold. For comparison, the 3T results were displayed at the same liberal threshold (Figure 2b). There was minimal evidence for selective activations at 1.5T, with one small exception in posterior cortex (Figure 2b; green; x,y,z = 0,-86,-12).

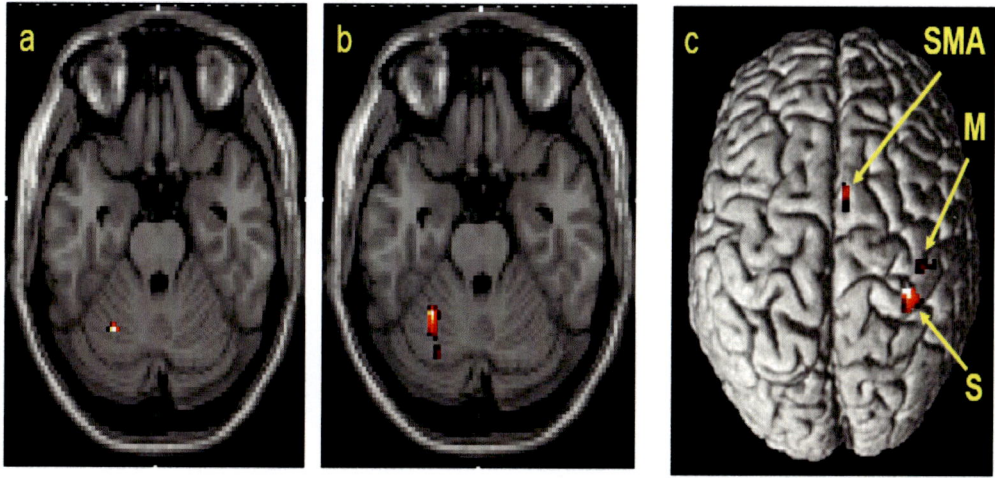

Figure 1. Activation of ipsilateral cerebellum at (a) 1.5T and (b) 3T and (c) contralateral sensory-motor areas at 3T. Images thresholded at p<0.05 (FWE corrected, k>10).

Activations at 3T were more extensive than at 1.5T (Figure 2b; red), but there was considerable overlap in terms of activated regions (Figure 2b; yellow). One apparent exception was observed sub-cortically, with selective activation at 3T of bilateral thalamus and the left globus pallidum / putamen region (Figure 3a). At a very low statistical threshold ($p < 0.05$ uncorrected), differential activation was seen at 1.5T in thalamus but not in pallidum/putamen (Figure 3b).

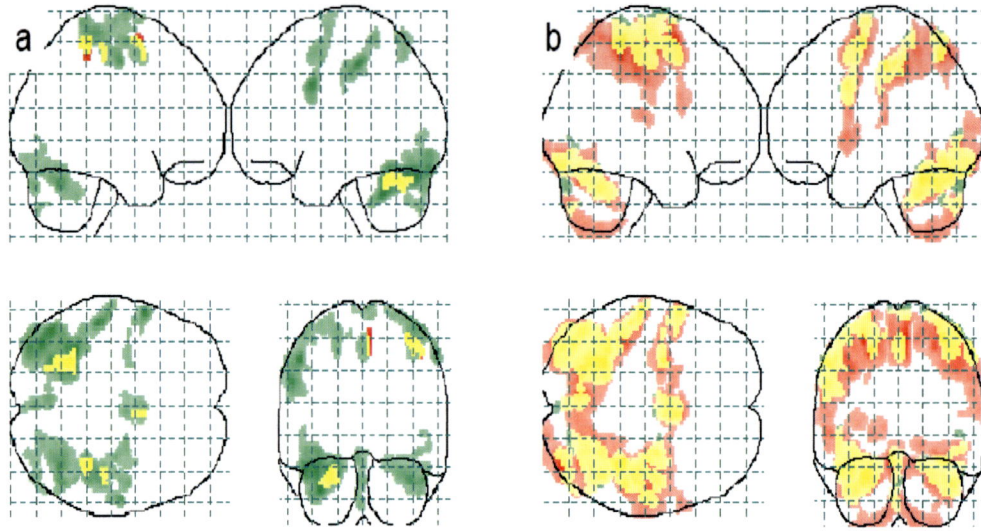

Figure 2. a) Activation patterns during finger tapping at 1.5T (green; threshold = 0.001 uncorrected) and 3T (red; threshold = 0.05 FWE corrected). Yellow = overlap. b) Activation patterns during finger tapping at 1.5T (green; threshold = 0.001 uncorrected) and 3T (red; threshold = 0.001 uncorrected). Yellow = overlap.

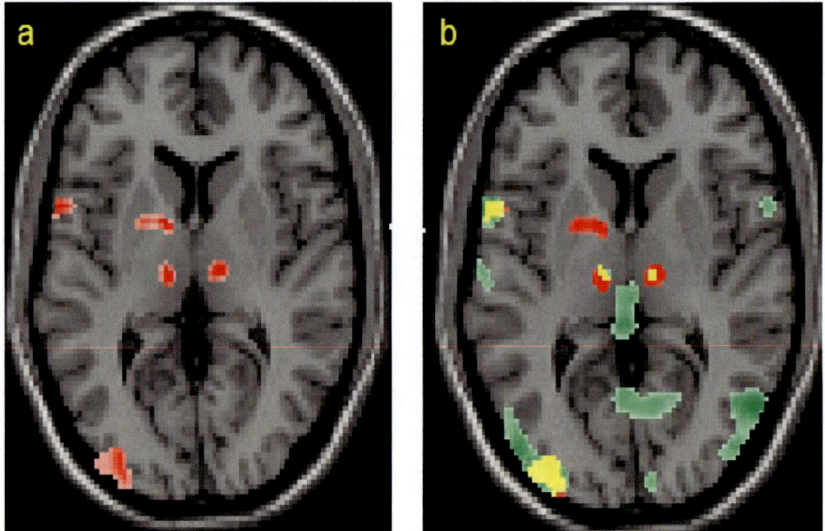

Figure 3. a) Activation of left putamen/pallidum (-12, 2, 6; -20,4,0, t = 6.30) and bilateral thalamus (+/- 12, -18, 6; t = 6.60/6.93) at 3T (threshold = p<0.001 uncorrected, k>10). b) Activation of left putamen/pallidum and bilateral thalamus at 3T (red; threshold = p<0.001 uncorrected, k>10) and at 1.5T (green; threshold = 0.05 uncorrected). Yellow = overlap.

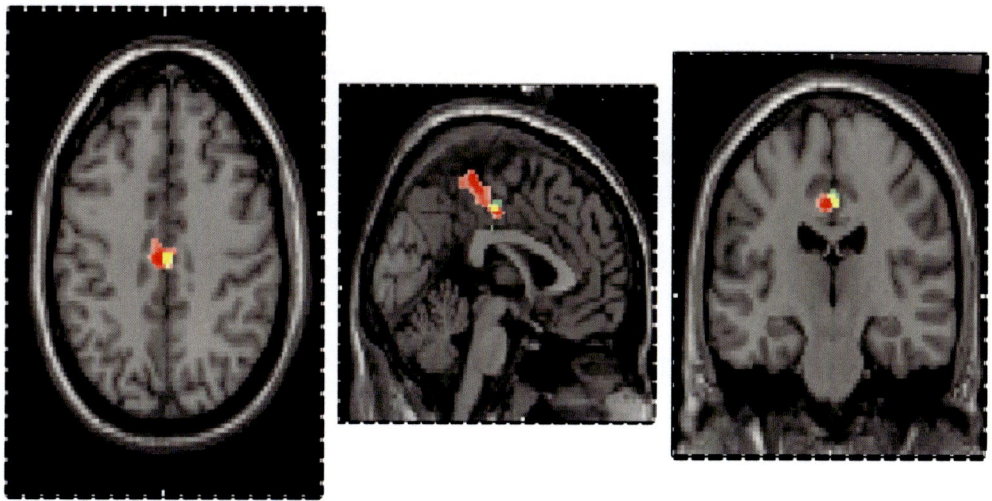

Figure 4. Medial motor cortex activation in comparison of trained vs. untrained finger tapping (threshold = 0.01 uncorrected). Red = 3T; Green = 1.5T; Yellow = overlap.

## Trained vs. Untrained Finger Tapping

At the most conservative statistical threshold, no areas of differential activity were detected at neither 1.5T nor at 3T. At the more liberal threshold, one region in the mid-cingulum region showed differential activation at 3T (x,y,z = -4,-20,40; $t$ = 5.20). At an even more liberal threshold (0.01 uncorrected) increased activity in this region was identified at 1.5T as well (Figure 4 green/yellow). At this threshold, the activated region was much more extensive at 3T (Figure 4, red), and did likely include parts of the SMA (see e.g., Riecker et al., 2003).

## Comparison of Cluster Extent at 1.5 and 3T

At 3T, in the comparison of tapping with rest, an extensive activation cluster was seen in the right hemisphere, involving sensory-motor and parietal areas (BA 2,3,4,5,6,7,40; k = 1468 at an threshold of $t > 7$). At the same cut-off for 1.5T, only two smaller clusters were revealed in the right cerebrum (BA 4; k = 6; BA 3; k =46). Several distinct activation clusters were revealed in right sensory and motor cortex when the 3T SPM-results were thresholded at a very stringent level ($t > 10$). However, at this level, no corresponding clusters at all could be detected at 1.5T. While this constitutes evidence that the activations were more extensive at 3T, it does not provide quantification of the difference. We therefore adjusted thresholds to levels at which distinct (i.e. spatially defined) clusters were revealed at both 1.5 and 3T and compared the associated cluster sizes (Table 1). Across the four selected clusters, at the same statistical threshold, the mean number of activated voxels was about 50% higher at 3T ($M$ = 121) than at 1.5T ($M$ = 62).

**Table 1. Extent of selected activations as a function of field strength**

| Area | MNI x, y, z coordinates | | Size of cluster | |
|---|---|---|---|---|
| | 1.5T | 3T | 1.5T | 3T |
| Left Cerebellum[1] | -24,-60,-24 | -26,-52,-24 | 48 | 118 |
| Left premotor cortex (area 4)[1] | -58,8,34 | -58,6,32 | 7 | 11 |
| Left parietal cortex (area 2)[1] | -50,-30,46 | -42,-38,48 | 6 | 26 |
| Supplementary Motor Area[2] | 2,-18,44 | -4,-20,40 | 186 | 330 |

## Results from Individual Subjects

To assess sensitivity differences between 1.5 and 3T at the level of individual subjects, selected peaks from the [finger tapping vs. rest] contrast were used as seed voxels. Figure 5 shows cerebellar activation for each individual subject at 3T (red) and 1.5T (green) along with overlapping activity at both field strengths (yellow). The left cerebellar region was identified at both field strengths in the random-effects analyses, and it was revealed for each individual subject at 3T and for the majority of subjects at 1.5T (no activation = subject # 3, 6, 12).

Figure 6 shows the same kind of results for right motor cortex. At the most stringent level, this region was only identified in the group analysis at 3T. At the individual level, right-motor cortex activation was seen for all 12 subjects at 3T and for seven of the 12 subjects at 1.5T (no activation = subject # 1, 3, 4, 6, 7).

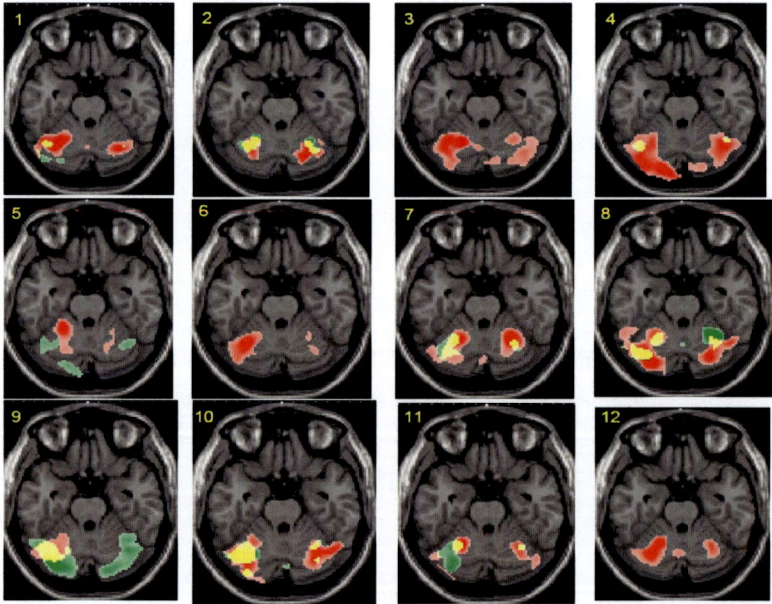

Figure 5. Individual activation patterns at 1.5T (green) and 3T (red) for left cerebellum (x,y,z = -24, -60, -24) with overlapping activity in yellow. Images were thresholded at p<0.01 (FWE corrected, k>10).

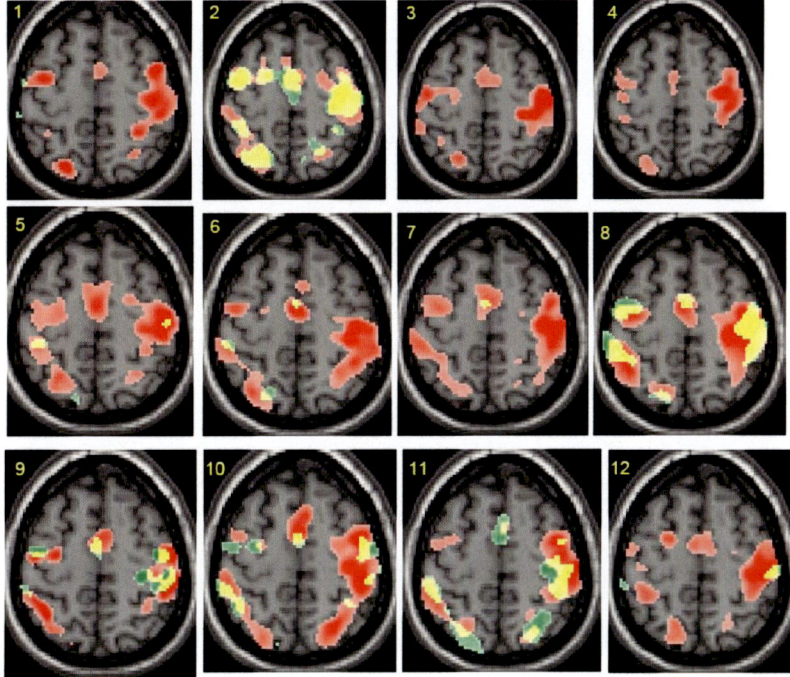

Figure 6. Individual activation patterns at 1.5T (green) and 3T (red) for right motor cortex (x,y,z = 42, -26, 56) with overlapping activity in yellow. Images were thresholded at p<0.01 (FWE corrected, k>10).

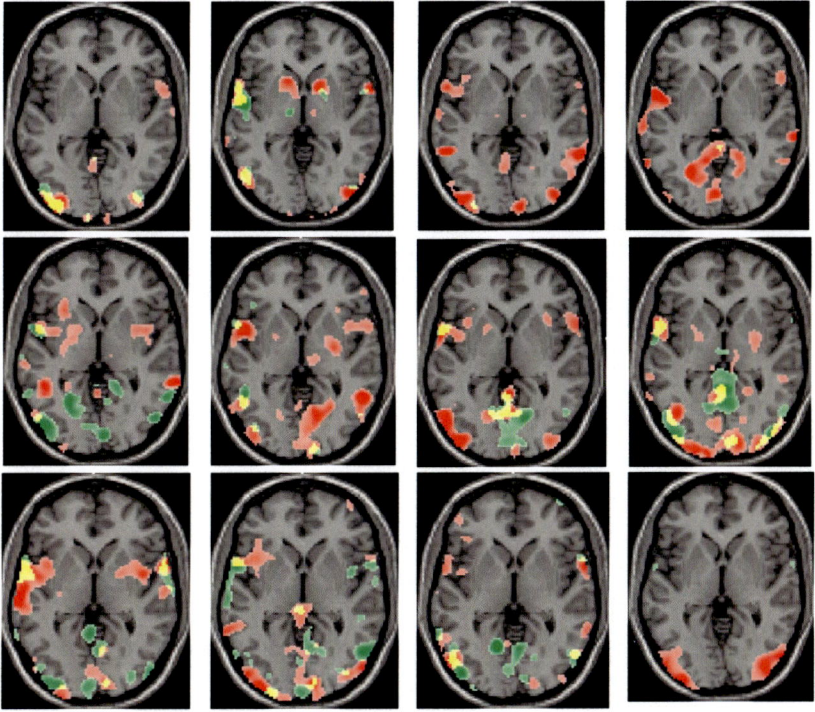

Figure 7. Individual activation patterns at 1.5T (green) and 3T (red) in basal ganglia (x,y,z = -20, 4, 0) with overlapping activity in yellow. Images were thresholded at p<0.001 (uncorrected, k>10).

Figure 7 shows individual activation patterns for subcortical regions (basal ganglia and thalamus). These regions displayed the most pronounced difference between 1.5T and 3T in the group analyses. Correspondingly, at the individual level, thalamus and basal ganglia activation was only detected for one subject at 1.5T (# 2). At 3T, thalamus activation was seen for 5 subjects (# 2, 3, 5, 6, 8) and left basal ganglia activation was seen for 6 subjects (# 2, 5, 6, 7, 8, 9).

## DISCUSSION

The findings of the present study reinforce the conclusion by Voss et al. (2006) that it is beneficial to conduct functional activation studies at 3T rather than at 1.5T. At the most stringent statistical level, ipsilateral cerebellar activity was seen at 1.5T in the comparison of tapping versus rest, whereas this region plus right sensory-motor cortex and SMA were revealed at 3T. Thus, in line with previous comparative studies (Nakai et al., 2001), our results suggest that 3T-imaging has better chances of revealing a whole functional network.

At a more liberal statistical level, 1.5T imaging identified the same regions to be engaged during finger tapping as were identified at the more conservative level with 3T imaging. Similarly, in the tighter comparison of trained and untrained finger tapping, the SMA, which has been associated with skilled finger tapping (e.g., Hikosaka et al., 1999), was detected at different statistical levels for both 1.5 and 3T. The pronounced overlap in activation for 1.5 and 3T when the threshold was adjusted suggests that claims of observing novel phenomena at higher field strengths should be treated with caution. At the same time, if lower field-strength imaging requires the use of more liberal statistical thresholds, then there is the obvious risk of false positives. In the present study there was little evidence that unique areas were revealed at 1.5T at a liberal threshold. Only a few small activations in posterior cortex were seen for 1.5T but not 3T, and these may constitute false positive activations.

For both the loose and tight comparisons, the extent of activations at a given statistical level was much more extensive at 3T than at 1.5T. This difference was very pronounced in the right sensory-motor cortex that was strongly engaged by the tapping task, but also substantial in other regions. In the direct comparison of selected clusters, the mean number of activated voxels was about 50% higher at 3T than at 1.5T. Thus, if one consider the very extensive activations at 3T that did not allow direct comparisons, these results suggest more than 50% increase in cluster size at 3T, which is in agreement with previous estimates (cf., Hoenig et al., 2005; Krasnow et al., 2003).

In addition to stronger and more extensive activations, the results provided some evidence for selective activations in bilateral thalamus and in the left striatum at 3T. Activation of these regions during finger tapping is to be expected (e.g., Riecker et al., 2003). Higher field-strength has previously been shown to be crucial for imaging of thalamic regions (Chen et al., 1999; Krainik et al., 2000), and Scholtz et al. (2000) found that percent signal changes in the basal ganglia were smaller by a factor of three than changes in the SMA and motor cortex. Consistent with the present findings, Scholtz et al. demonstrated that regardless of the hand used, the left basal ganglia were more active than the right in right-handed subjects.

We also evaluated the results at the individual level and found a much more consistent between-subject pattern at 3T than at 1.5T. Particularly for subcortical regions, the individual patterns reinforced the conclusion that 3T imaging is more sensitive than 1.5T imaging. It should be noted that 1.5T imaging of individuals with brain tumors has been successfully done pre-surgically (e.g., Håberg et al., 2004), but it has also been noted that 3T scanning should offer improved sensitivity (Vlieger et al., 2004). Our findings support that expectation. Still, it is important to stress that we only noted basal ganglia activation in half of our subjects, which is in line with the comment by Scholz et al. (2000) that there is still no task which reliably detects basal ganglia activation on a single subject basis. We used an event-related design; future 3T studies with blocked designs may have better chances of revealing subcortical activations in individual subjects.

Obviously, other factors than field strength can have influenced the observed differences between 1.5T and 3.0T imaging. The possibility to use shorter repetition time on the 3T scanner, with preserved spatial resolution, means increased temporal resolution of the fMRI data. In the present case, for a specific time period, the number of data points was a factor of 1.5 higher for the 3T data. Another difference was type of coil. The 8-channel SENSE head coil used in the 3T scanner has increased signal to noise ratio (SNR) compared to the quadrature head coil used in the 1.5T scanner, especially in peripheral cortical regions (de Zwart et al. 2002a). The use of SENSE in the 3T makes it possible to increase both temporal and spatial resolution (Pruessmann et al. 1999), here in the z direction, and is also advantageous from a susceptibility artifacts point of view (de Zwart et al. 2002b). The drawback is the degradation in image SNR, and these benefits and drawbacks should preferably be balanced taking the expected activated areas into account (Schmidt et al. 2005). With these caveats concerning the comparison in mind, we conclude that functional imaging at 3T is beneficial.

## REFERENCES

Chen, W., Zhu, X.-H., Thulborn, K. R. and Ugurbil, K. (1999). Retinotopic mapping of lateral geniculate nucleus in humans using functional magnetic resonance imaging. *PNAS, 96*, 2430-2434.

Fera, F., Yongbi, M. N., van Gelderen, P., Frank, J. A., Mattay, V. S., and Duyn, J. H. (2004). EPI-BOLD fMRI of human motor cortex at 1.5T and 3.0T: Sensitivity dependence on echo time and acquisition bandwidth. *Journal of Magnetic Resonance Imaging, 19*, 19-26.

Håberg, A., Kvistad, K. A., Unsgård, G., and Haraldseth, O. (2004). Preoperative blood oxygen level-dependent functional magnetic resonance imaging in patients with primary brain tumors: Clinical application and outcome. *Neurosurgery, 54*, 902-915.

Hikosaka, O., Nakahara, H., Rand, M. K., Sakai, K., Lu, X., Nakamura, K., et al. (1999). Parallel neural networks for learning sequential procedures. *TINS, 22*, 464-471.

Hoenig, K., Kuhl, C. K., and Scheef, L. (2005). Functional 3.0-T MR assessment of higher cognitive function: Are there advantages over 1.5-T imaging? *Radiology, 234*, 860-868.

Krasnow, B., Tamm, L., Greicius, M. D., Yang, T. T., Glover, G. H., Reiss, A. L., and Menon, V. (2003). Comparison of fMRI activation at 3 and 1.5 T during perceptual, cognitive, and affective processing. *NeuroImage, 18*, 813-826.

Krainik, A., Lehéricy, S., Hennel, F., van de Moortele, P.-F., Marsault, C., and Le Bihan, D. (2000). Comparison of cortical and basal ganglia activation at 1.5T and 3T. *NeuroImage, 11*, S522.

Krüger, G., Kastrup, A., and Glover, G. H. (2001). Neuroimaging at 1.5 T and 3.0 T: Comparison of oxygenation-sensitive magnetic resonance imaging. *Magnetic Resonance in Medicine, 45*, 595-604.

Nakai, T., Matsuo, K., Kato, C., Okada, T., Moriya, T., Isoda, H., Takehara, Y., and Sakahara, H. (2001). BOLD contrast on a 3 T magnet: Detectability of the motor areas. *Journal of Computer Assisted Tomography, 25*, 436-445.

Nyberg, L., Eriksson, J., Larsson, A., and Marklund, P. (2006). Learning by doing versus learning by thinking: An fMRI study of motor and mental training. *Neuropsychologia, 44*, 711-717.

Pruessmann K. P., Weiger M., Scheidegger M. B., and Boesinger P. 1999. SENSE: sensitivity encoding for fast MRI. *Magn. Reson. Med. 42,* 952-962.

Riecker, A., Wildgruber, D., Mathiak, K., Grodd, W., and Ackermann, H. (2003). Parametric analysis of rate-dependent hemodynamic response functions of cortical and subcortical brain structures during auditorily cued finger tapping: a fMRI study. *NeuroImage, 18*, 731-739.

Rorden, C. and Brett, M. (2000). Stereotaxic display of brain lesions. *Behavioural Neurology, 12*, 191-200.

Schmidt C. F., Degonda N., Luechinger R., Henke K., and Boesinger P. 2005. Sensitivity-encoded (SENSE) echo planar fMRI at 3T in the medial temporal lobe. *NeuroImage 25*, 625-641.

Scholtz, V. H., Flaherty, A. W., Kraft, E., Keltner, J. R., Kwong, K. K., Chen, Y. I., Rosen, B. R., and Jenkins, B. G. (2000). Laterality, somatotopy and reproducibility of the basal ganglia and motor cortex during motor tasks. *Brain Research, 879*, 204-215.

Ugurbil, K., Hu, X., Chen, W., Zhu, X.-H., Kim, S.-G., and Georgopoulos, A. (1999). Functional mapping in the human brain using high magnetic fields. *Phil. Trans. R. Soc. Lond. B, 354*, 1195-1213.

Vlieger, E.-J., Majoie, C. B., Leenstra, S., and den Heeten (2004). Functional magnetic resonance imaging for neurological planning in neurooncology. *Eur. Radiol. 14*, 1143-1153.

Voss, H. U., Zevin, J. D., and McCandliss, B. D. (2006). Functional MR imaging at 3.0 T versus 1.5 T: A practical review. *Neuroim. Clin. N. Am., 16*, 285-297.

de Zwart J. A., Ledden P. J., Kellman P., van Gelderen P., and Duyn J. H. 2002a. Design of a SENSE-optimized high-sensitivity MRI receive coil for brain imaging. *Magn. Reson. Med. 47*, 1218-1227.

de Zwart J. A., van Gelderen P., Kellman P., and Duyn J. H. 2002b. Application of sensitivity-encoded echo-planar imaging for blood oxygen level-dependent functional brain imaging. *Magn. Reson. Med. 48*, 1011-1020.

# INDEX

## A

accidents, 148
accuracy, viii, ix, 41, 42, 43, 46, 48, 49, 51, 58, 59, 63, 76, 77, 79, 84, 95, 111, 112, 113, 130, 153, 155, 164
acid, 73, 76, 86, 89, 146
acidosis, 158
acoustic neuroma, 115, 116, 126
acquisitions, viii, 31, 38, 42, 44, 163
acrylate, 76
activation, vii, viii, x, 2, 7, 8, 11, 12, 14, 16, 17, 19, 20, 21, 22, 23, 24, 26, 27, 28, 29, 47, 49, 52, 54, 57, 58, 96, 98, 99, 102, 105, 106, 107, 109, 152, 159, 161, 162, 163, 165, 167, 168, 169, 170, 171, 172
acute glaucoma, 118
acute ischemic stroke, 2
adaptation, 146
additives, 82
adenosine, 2, 129, 140
adenosine triphosphate (ATP), 129, 140
adhesion, 71, 73, 74, 86, 156
adipose tissue, 80, 81, 90
adjustment, 76, 78
administration, 142
adolescence, 94
adolescent patients, 157
adolescents, 157
adult stem cells, 81, 89, 90
adulthood, 94
adults, 155, 159
advertisements, 155
affective disorder, 101
age, 32, 133, 141, 143, 154, 162
agent, 34
aggression, 159
aggressiveness, 159
aging, 105, 154, 156
alcohol, 150
algorithm, 46, 54, 55, 66, 87, 109, 134
allele, 102
alternative, 5, 50, 51, 53, 70, 81
Alzheimer's disease (AD), ix, x, 86, 97, 105, 129, 130, 131, 139, 140, 145, 153, 154, 155
American Psychiatric Association, 158
amine, 131, 157, 159
amplitude, 43, 44, 133
Amsterdam, 66, 67
amygdala, 9, 16, 54, 56, 131, 140, 147, 149, 151, 152, 160
anaerobic, 146
analgesic, 26
anastomosis, 84
anatomy, 43, 49, 50, 61, 102, 126, 133, 147
aneurysm, 118, 119
angiogenesis, 91
angiography, viii, 31, 32, 38, 39, 46, 125, 127
animal models, 2, 68, 82, 102, 112
animal welfare, 25
animals, 8, 25, 68
ANOVA, 129, 136, 137
anterior cingulate cortex, 9, 96, 106, 158
anticholinergic effect, 109
anticonvulsant, 140
antidepressant(s), 141, 142, 145
antinociception, 24
antrum, 116
anxiety, 19, 28, 131, 145, 152, 156, 158, 159, 160
anxiety disorder, 131, 152
apoplexy, 118
apoptosis, 1
arousal, 146, 152
arteriovenous malformation, 39, 117, 127
arteritis, 118

artery(ies), viii, 3, 4, 31, 32, 33, 34, 35, 37, 39, 42, 43, 46, 112, 113, 114, 115, 118, 119, 126, 128
arthritis, 21, 27
aspartate, 151
assessment, x, 3, 26, 32, 39, 94, 95, 100, 104, 113, 125, 127, 130, 131, 132, 171
assignment, 163
assumptions, 49, 98
asymmetry, 104, 134, 148
asymptomatic, 3, 114
atherosclerosis, 3
atherosclerotic plaque, 3
atrophy, 113, 114
attachment, 73
attacks, 123, 160
attention, 22, 94, 102, 147
attitudes, 159
atypical, 21, 27, 117, 123, 124, 142, 144, 157
auditory cortex, 96
autism, 97
automatization, 79
autoradiography, vii, 7, 11, 12, 16, 17, 20, 23, 24, 25, 68
availability, 81, 131
averaging, 154
aversion, 19
avoidance, 146
axons, 112

## B

back pain, 14, 23, 29
bacteria, 118
bandwidth, 33, 34, 72, 171
barriers, 107
basal forebrain, 152
basal ganglia, x, 9, 99, 107, 148, 156, 161, 169, 170, 171, 172
basement membrane, 6
basilar artery, 115
behavior, viii, 5, 26, 28, 31, 32, 33, 90, 101, 151
Belgium, 72
beliefs, 123
Bell's palsy, 120, 127
beneficial effect, 3
benefits, 3, 171
benign, 115
biocompatibility, 73, 75, 84
biodegradable, 86, 87, 88
biogenesis, 156
biological activity, 86
biological interactions, 5
biological markers, ix, 93, 94, 100

biological responses, 5
biomarker(s), 94, 95, 97, 100, 109
biomaterials, 5, 74, 79, 82, 86
biomedical, 5, 65, 86
biomedical applications, 65
biomolecules, 74, 77, 78, 79
bioreactor(s), 71, 82, 84, 89, 90
biotechnology, 5
bipolar disorder, 94, 101, 104, 108
birth, 81, 104
bladder, 82, 91
blood, x, 4, 20, 21, 28, 32, 33, 34, 35, 80, 81, 89, 90, 101, 104, 105, 112, 125, 129, 130, 131, 135, 147, 153, 154, 155, 156, 157, 158, 159, 160, 171, 172
blood flow, x, 4, 20, 28, 105, 112, 129, 130, 131, 135, 147, 153, 154, 155, 156, 157, 158, 159, 160
blood sampling, 131
blood supply, 4
blood vessels, 80, 81
blurring, viii, 41, 42, 46
BMP-2, 82
BOLD, x, 21, 23, 26, 96, 105, 161, 171, 172
bonding, 76
bone, 5, 71, 72, 73, 75, 80, 81, 82, 83, 85, 87, 89, 90, 91, 117, 120
bone grafts, 91
bone marrow, 80, 81, 82, 89, 90
bone marrow transplant, 89
bone morphogenetic proteins, 91
bounds, 60
bowel, 29, 153
bradycardia, ix, 109
brain activity, vii, 7, 8, 11, 19, 20, 27, 28, 60, 96, 98, 105, 159, 164
brain development, 94, 103
brain imaging techniques, 100
brain stem, 16, 117, 121, 123
brain structure, vii, 7, 10, 11, 12, 14, 16, 17, 18, 19, 21, 98, 102, 131, 150, 152, 172
brain tumor, 68, 171
brainstem, 9, 16, 23, 27, 29, 125, 152
breakdown, 100
bronchial tree, 122
buccal mucosa, 88
bulk materials, 76, 77
burning, 21, 27, 80, 112
buttons, 164

## C

calibration, 72
calvarium, 71, 72
Canada, 31, 39, 41, 164

cancer, 122, 127
candidates, 2, 62, 63, 88
carbon, 75, 131
carbon dioxide, 75
carcinoma, 116, 119, 121, 122, 126
cardiac risk, 4
caregiver, x, 130
carotid arteries, 34, 39
carotid endarterectomy, 3, 4
cartilage, ix, 69, 71, 72, 73, 74, 80, 81, 82, 83, 87, 88, 90, 91
cartilaginous, 84, 85
cast, 87, 123
causal model, 98, 107
causal relationship, 140
causalgia, 89
cell, 1, 5, 70, 73, 74, 79, 80, 81, 82, 84, 85, 86, 89, 90, 119, 135, 145, 156, 158
cell adhesion, 71, 86, 156
cell culture, 6, 80
cellulose, 78
central nervous system (CNS), 8, 10, 13, 25, 30, 155
cephalgia, 128
ceramic(s), 73, 75, 76, 77, 78, 79, 85, 87
cerebellopontine angle tumor, 126
cerebellum, x, 9, 16, 99, 134, 139, 151, 161, 165, 168
cerebral arteries, 33
cerebral blood flow (CBF), x, 2, 4, 20, 28, 105, 129, 130, 131, 132, 133, 136, 137, 146, 147, 150, 152, 153, 154, 156, 157, 158, 159
cerebral hypoperfusion, 4
cerebrum, x, 161, 165, 167
certainty, 110
cervical, 39, 122, 123
channels, 32, 51, 55, 77
chemical, vii, 5, 8, 25, 26, 78, 90, 115
chemical properties, 90
childhood, 159
childhood sexual abuse, 159
children, 53, 57, 150, 156
chondrocyte(s), 71, 80, 81, 82, 83, 85, 86
chromosomes, 140
chronic disorders, 18, 21, 23
chronic fatigue syndrome, 156
chronic pain, 17, 19, 21, 22, 24, 27, 124, 126
circulation, 4
citalopram, 149, 152
classes, 49
classification, 66, 96, 104, 123, 137, 139
clinical diagnosis, x, 35, 114, 130, 131, 137
clinical examination, 124
clinical neurophysiology, 62

clinical presentation, 94, 111, 115, 123, 124, 126
clinical syndrome, 25, 102
clinical trials, 2, 82, 84
clinicians, 13
clinics, 21, 25, 100
closure, 3
cluster analysis, 55
cluster headache, 123
clustering, 66, 137
clusters, x, 136, 137, 138, 161, 167, 170
coagulation, 88, 114
coding, 20, 28
cognition, 62, 98, 101, 106
cognitive, viii, x, 8, 19, 20, 21, 22, 23, 24, 28, 61, 94, 96, 97, 98, 99, 100, 102, 108, 130, 149, 151, 152, 156, 171
cognitive domains, 98, 100
cognitive function, 102, 151, 171
cognitive impairment, x, 130
cognitive level, 151
cognitive process, 97, 108
cognitive processing, 97
cognitive restructuring therapy, 152
cognitive tasks, 61, 96
coherence, 66
cohort, 95, 96, 102, 104
collaboration, 2, 101, 107
collagen, 77, 82, 83, 88, 89, 90
collateral, 2, 4
commissure, 89
communication, 101, 151
community, x, 97, 130, 155
comorbidity, 155
compensation, 113, 146
complex partial seizure, 67
complexity, ix, 44, 65, 69, 74, 95, 96, 110
complications, 3, 4, 109, 118, 127
components, 3, 5, 9, 57, 65, 75, 76, 96, 99, 131, 136, 137
composite, 85
composites, ix, 69, 74, 78
composition, 78, 85
compounds, 8, 76
computation, 44, 46, 60
computed tomography (CT), vii, 63, 71, 83, 87, 116, 117, 118, 119, 122, 123, 128, 130, 154, 157, 159, 160
computer added design (CAD), ix, 69, 70, 72, 74, 75, 78, 79, 87
computerized brain atlas (CBA), 129, 133, 141, 142, 143, 144
computers, 71
computing, 132

concentrates, x, 131
concentration, 38
conditioning, 151
conduction, 63
conductivity, 49
conductor, 61
confidence, 38, 112
configuration, 32, 52, 55
conflict, 126
confounding variables, 2
Connecticut, 93
connective tissue, 80
connectivity, 95, 97, 98, 99, 102, 105, 106, 137, 138, 155
consensus, x, 59, 110, 130, 153
consent, 32, 162
construction, 107
control, 1, 2, 9, 21, 28, 82, 90, 95, 100, 101, 104, 113, 114, 132, 133, 136, 141, 143, 148, 150
control group, 143, 148, 150
convergence, 108
conversion, 135
cornea, 80
corpus callosum, 57
correlation analysis, vii, 7, 12, 17
correlation coefficient, 24, 46
correlation(s), vii, 2, 7, 12, 17, 20, 24, 45, 46, 63, 94, 95, 97, 98, 99, 123, 125, 126, 138, 153, 156
cortex, 9, 16, 18, 19, 20, 21, 22, 24, 27, 29, 52, 54, 57, 64, 96, 104, 106, 107, 141, 142, 143, 147, 148, 149, 151, 152, 158, 159, 160, 162, 165, 167, 168, 169, 170, 171, 172
cortical systems, 94
cortisol, 151
costs, 4, 45, 76, 77
coupling, 94, 132
coverage, 43, 51, 52
covering, 44
cranial nerve, ix, 111, 121, 122, 123, 124, 125, 127
craniopharyngioma, 118
craniotomy, 114
cues, 146
cultivation, 82
culture, 5, 80, 82, 84
current limit, 87
cybernetics, 66
cycles, 43, 44, 45
cytoarchitecture, 94
cytochrome, 145
cytotoxic, 71

# D

data analysis, 96, 135
data processing, 71, 79
data set, 96, 106
database, 100, 132, 133
death, 3, 139
decisions, 162
decomposition, 99, 154
deconvolution, 105
defects, 70, 72, 73, 80, 87, 91, 160
deficiency, 140
deficit(s), ix, 2, 98, 102, 107, 110, 111, 112, 154
definition, 49, 97
deformation, 5
degradation, 38, 71, 73, 77, 79, 87, 171
delivery, 73
demand, 28
dementia, ix, 129, 130, 153
demyelination, 112
Denmark, 111
density, 51, 54, 60, 62, 65, 104, 134, 139
dependent variable, 137
deposition, 70, 74, 78, 85, 87, 91
depression, x, 94, 130, 139, 140, 141, 142, 143, 144, 145, 146, 152, 153, 155, 156, 157, 158, 160
depressive symptoms, 155
dermatomyositis, 118
desensitization, 149, 152, 158
detection, ix, 13, 34, 66, 109, 114, 125, 137, 153, 155
developmental origins, 104
deviation, 34, 43, 45
diabetes, 142
diagnostic criteria, 113, 114, 124
differential diagnosis, 126, 153
differentiated cells, 71, 80, 81
differentiation, 6, 82, 90
diffuse optical tomography, vii
diffusion, 70
dimer, 131
diplopia, 118
dipole, 49, 53, 54, 55, 56, 60, 61, 62, 63, 64, 65
direct measure, 4
disability, 141, 155
discharges, 48, 54, 57, 58, 64, 68
discriminant analysis, 65, 104, 129
discrimination, x, 62, 130, 155
disorder, x, 21, 27, 94, 95, 98, 102, 103, 104, 108, 118, 123, 130, 140, 142, 146, 147, 150, 156, 157, 158, 159, 160
displacement, 43, 45, 114, 115
distress, x, 130, 139

distribution, x, 49, 50, 51, 55, 73, 112, 123, 130, 132, 133, 134, 136, 137, 138, 139, 150, 151, 160
diversity, 98
division, 122
DNA, 88, 101, 130, 140, 155, 156
dopamine, 140, 155
dorsal horn, 9
dorsolateral prefrontal cortex, 29, 107
drug discovery, 6, 8, 26
drug safety, 155
drugs, 8, 10, 26, 73, 140
DSM, 102, 146, 158
DSM-II, 146
DSM-III, 146
DSM-IV, 158
duration, 2, 24, 43, 118, 164
duties, 148, 149
dysarthria, 117, 121

# E

education, 87
EEG, vii, viii, 7, 8, 11, 17, 47, 48, 49, 50, 51, 53, 54, 55, 56, 57, 59, 60, 61, 62, 63, 64, 65, 66, 67, 98, 100, 103, 107
EEG activity, 61
EEG patterns, 48, 55, 65, 66
elasticity, 71
elderly, 141, 157, 160
electric field, 51, 52
electric potential, 65
Electric Source Imaging (ESI), viii, 47, 48, 49, 50, 51, 52, 53, 54, 55, 58
electrical fields, 62
electrodes, viii, 47, 48, 49, 51, 52, 53, 54, 55, 57, 58, 61, 62, 66
electroencephalography, 59, 63, 64, 67, 154
electrolyte, 1
electromagnetic, vii, 49, 60, 62, 63, 64
electromagnetic tomography, vii
embryo, 81
embryonic, 81
embryonic stem cells, 81
emission, ix, 27, 28, 29, 63, 97, 130, 156, 157, 159, 160
emitters, 131
emotion, 28
emotional experience, 8
emotional responses, 20, 152
emotions, 9, 19, 24, 30, 147, 151, 159
encapsulation, 82
encoding, 32, 39, 151, 172
endometrial carcinoma, 116

endophenotypes, ix, 93, 101, 102, 108
endoscopy, 123, 128
energy, 1, 76, 78, 133, 134, 140, 156
engagement, 96
England, 34
entorhinal cortex, 9, 16, 19
environment, ix, 6, 69, 73, 78, 80, 90, 93, 132
enzyme(s), 145, 146
epidemiology, 158
epigenetic, 101
epilepsy, viii, 47, 48, 51, 53, 55, 58, 59, 60, 61, 62, 63, 64, 65, 66, 67, 68, 140, 155
epileptic seizures, 54, 56, 65, 66
epileptogenesis, 68
episodic memory, 107
epithelium, 81
equipment, 72, 74, 79
esophagus, 123
estimating, viii, 41, 42, 43, 44, 45, 49
ethics, 162
ethylene glycol, 73
etiology, 101, 127
evidence, 1, 3, 6, 17, 18, 27, 30, 66, 94, 98, 102, 105, 110, 112, 114, 120, 140, 153, 156, 157, 158, 165, 167, 170
evoked potential, 29
evolution, 2, 8, 65, 84, 145
examinations, 48, 131
excision, 72
exclusion, 123
execution, 151, 152
exertion, 142
exposure, 20, 141, 147, 149, 152, 158, 159
extinction, 151
extracellular, 6, 73, 135
extracellular matrix, 6, 73
extraction, 104
extraocular muscles, 118
extrusion, 78
eye movement, 152, 158

# F

fabrication, 74, 76, 78, 79, 81, 84, 85, 87, 88
facial nerve, 120, 122, 127
facial pain, ix, 21, 27, 111, 115, 116, 118, 119, 120, 121, 122, 123, 124, 126, 127, 128
facial palsy, 117
failure, 2, 142
false alarms, 137
false positive, 137, 170
family, 81
fat, 71, 81, 119, 156

fatigue, 139, 156
fatty acids, 141, 156
fear, 146, 147, 151, 160
fear response, 146
feelings, 139, 146
females, 141
FFT, 55, 56, 57, 65
fiber(s), 12, 13, 29, 77, 78, 85, 91, 98, 158
fibrin, 73, 74, 82, 86
fibromyalgia, 14, 23, 27
filament, 75, 77
fillers, 76
financial support, 25
finite element method, 50
flashbacks, 146
flight, 39, 125
fluid, 90
fluoxetine, 155
focusing, ix, 111
follicle, 81
forebrain, 18, 20, 28, 152
Fourier, 45, 125
Fourier transformation, 125
frontal cortex, 16, 18, 20, 142, 143, 147, 149, 159
frontal lobe, 19, 20, 50, 54, 55, 56, 58, 64, 108, 129, 141, 142, 143, 144, 145, 146, 148, 149, 156
Frontal Lobe Dementia (FLD), ix, x, 129, 130, 131
fumarate, 76
functional activation, 102, 170
functional changes, 98, 153
functional imaging, vii, 7, 8, 11, 18, 25, 27, 32, 52, 95, 96, 105, 135, 145, 155, 171
functional MRI (fMRI), vii, x, 7, 8, 10, 11, 17, 20, 21, 22, 23, 24, 25, 26, 27, 28, 29, 30, 52, 59, 94, 95, 96, 97, 98, 99, 100, 102, 105, 106, 107, 129, 150, 161, 162, 163, 164, 171, 172
fungus, 118
fusion, 75, 78, 94, 97, 98

## G

gadolinium, 34, 39, 113, 116, 120
ganglion, ix, 22, 29, 109, 112, 114, 118, 126
gastrointestinal tract, 80
gender, 100, 143, 154
gender differences, 154
gene expression, 104, 108
generation, 66
gene(s), 101, 102, 104, 108, 140, 156
genetic testing, 101
genetics, ix, 93, 94, 95, 101, 102, 103, 108
Geneva, 47, 59
genome, 101

genomics, 102
genotype, 81, 102, 108
Germany, 1, 5, 7, 69, 106, 109
gland, 118, 123
glass transition temperature, 79
glass(es), 5, 73, 75, 79
glaucoma, 118
glial, 145, 158
glial cells, 158
globus, 143, 151, 165
glossopharyngeal nerve, 117, 122
glucose, 2, 130, 132, 158
glucose metabolism, 158
glutamate, 151
glutathione, 130, 135, 154
glycol, 73
goals, 59, 100
gold, 42, 46, 137
gravity, 51, 78
grey matter, 49, 50, 102, 132, 139
groups, 24, 73, 136, 137, 150, 154, 155
growth, 6, 71, 73, 74, 82, 91, 104
growth factors, 71, 73, 74, 82, 91
guidance, 73, 89, 109
guilt, 139
guilt feelings, 139
gyrus, 20, 21, 54, 56, 95, 142, 149, 151

## H

habituation, 18, 151
hair follicle, 81
hands, 25, 48
head, viii, 6, 32, 33, 35, 37, 38, 39, 47, 49, 50, 51, 52, 58, 60, 61, 62, 71, 72, 75, 76, 78, 80, 82, 86, 88, 89, 134, 163, 171
headache, 123, 124
health, 5, 26, 29, 155
heart, 42, 43, 44, 45, 46, 80, 81, 109, 135, 151, 160
heart disease, 81
heart rate, 151, 160
heat, vii, 7, 12, 13, 19, 20, 22, 27, 28, 29, 78
heating, 75
hemisphere, 4, 95, 121, 148, 150, 151, 167
hemodynamic effect, 4
hemodynamics, 4, 107, 158
hemorrhage, 127
hepatocyte growth factor, 91
hepatocytes, 81
herpes zoster, 120, 127
heterogeneity, 18
high resolution, viii, 31, 38, 113

hippocampus, 9, 54, 56, 97, 102, 104, 140, 147, 148, 149, 151, 153
histogram, 99
HIV, 118
homeostasis, 1
homogeneity, 32
hormone, 156
host, 6, 73
host tissue, 6
hostility, 151
human brain, ix, 18, 23, 25, 27, 28, 48, 60, 62, 93, 94, 106, 132, 153, 172
human genome, 101
hunting, 102
hydrogels, 76, 78, 82, 85, 89
hydrophilic, 131, 135
hydrophobic, 82
hyperactivity, 27, 109, 151
hyperalgesia, vii, viii, 7, 8, 10, 11, 12, 13, 14, 16, 17, 18, 19, 20, 21, 22, 23, 24, 25, 26, 27, 28, 29
hyperarousal, 159
hyperesthesia, 29
hyperreactivity, 152
hypersensitivity, 13, 29, 152
hypoperfusion, 4
hypotension, ix, 109
hypothalamus, 9, 16, 28
hypothesis, 18, 24, 94, 95, 103, 104, 112, 113, 124, 135, 137, 140, 141, 151, 155, 160
hypothyroidism, 157
hypoxia, 68

## I

iatrogenic, 70
identification, ix, x, 2, 48, 66, 93, 101, 102, 130, 132
idiopathic, 23, 29, 118, 123, 126, 139
idiosyncratic, 112
image analysis, ix, x, 34, 93, 95, 130
imagery, 53, 158, 159
images, viii, x, 2, 31, 33, 34, 35, 36, 38, 39, 42, 43, 52, 61, 71, 72, 94, 98, 104, 107, 112, 130, 133, 136, 151, 161, 163, 164
imaging modalities, ix, 12, 17, 93, 100, 109
imaging techniques, 32, 48, 52, 74, 100, 125
immobilization, 86
immune reaction, 82
immune response, 84
immunocompromised, 118
immunological, 73
implants, ix, 69, 70, 71, 74, 84, 87
implementation, 82, 132
impulsive, 159

in situ, 89
in utero, 94
in vitro, 5, 81, 88, 154
in vivo, 6, 87, 88, 91, 104, 147
incidence, 12, 17, 115, 116
inclusion, 132, 139, 142
independent variable, 137
indication, 52
individual differences, 51, 136
induction, 82, 151
industrial production, 74
infancy, 84
infarction, 117, 121, 126, 127, 128
infection, 118, 120, 123
infectious diseases, 118
inflammation, 16, 118, 120
inflammatory, 3, 10, 16, 118, 121
informed consent, 162
infrastructure, 100
inheritance, 156
inhibition, 151, 152
inhibitor(s), 140, 152, 160
inhibitory, 9, 23, 56, 140, 145
inhibitory effect, 140
initiation, 55
injury, 1, 14, 16, 29, 114
innovation, 80, 122
insight, 13, 102
instability, 110
institutions, 100
insulin, 86
integration, ix, 24, 90, 93, 94, 95, 97, 138
intensity, 14, 20, 21, 33, 34, 35, 36, 112, 134
interaction(s), 5, 42, 86, 136, 137, 138, 154
interdependence, 103
interference, 113, 125
International Association for the Study of Pain (IASP), 8, 12, 26, 124
interpretation, 51, 54, 61, 134, 139
interval, 43, 163
intervention, 1, 3, 94, 123, 124, 125
intracerebral, 54, 64, 66
intracranial arteries, viii, 31, 32, 38
intracranial tumors, 126
intravital microscopy, vii
investment, 131
ipsilateral, x, 22, 23, 24, 123, 161, 165, 170
Ireland, 41
irritability, 146
irritable bowel syndrome, 29
ischemia, 1, 2
ischemic stroke, 1, 2
isolation, 117

Italy, 129

## J

joints, 71
justification, 13

## K

keratinocytes, 80, 81
kernel, 164
kidney, 80, 81
killing, 148, 149

## L

labeling, 4
labor, 51
lactic acid, 146, 158
language, vii, 7, 11, 147
larynx, 123
laser, 29, 74, 75, 76, 87
leaching, 88
lead, 3, 8, 38, 49, 63, 71, 82, 102
learning, 106, 146, 152, 160, 165, 171, 172
left hemisphere, 151
lesions, 13, 22, 50, 56, 61, 67, 115, 117, 118, 120, 123, 124, 160, 172
leukemia, 81
Lewy's Bodies, x, 130
life sciences, 79
lifetime, 139, 143, 144, 147
ligament, 117
likelihood, 140
limbic system, 9, 142, 147, 151
limitation, 98, 137
linear model, 164
linkage, 101
links, x, 130
lipophilic, 131, 135
literature, viii, 8, 11, 13, 18, 33, 47, 57, 98, 113, 135, 147, 155
liver, 80, 81, 88, 135
localization, viii, 47, 48, 51, 53, 54, 55, 56, 58, 59, 60, 61, 62, 63, 65, 67, 73
location, 16, 49, 53, 56, 61, 152
London, 63, 111, 164
longitudinal studies, 153
low back pain, 14, 23, 29
lung cancer, 122
lying, 18
lymphatic, 81

## M

magnet, 172
magnetic field, 22, 32, 59, 65, 172
magnetic resonance, viii, 4, 26, 27, 29, 31, 32, 39, 42, 46, 60, 61, 63, 70, 95, 103, 104, 106, 107, 112, 125, 129, 140, 156, 171, 172
magnetic resonance imaging (MRI), vii, viii, 4, 26, 27, 28, 29, 30, 33, 39, 46, 47, 48, 49, 50, 51, 52, 58, 60, 62, 63, 64, 94, 95, 100, 103, 104, 105, 106, 107, 110, 112, 114, 117, 118, 119, 120, 122, 123, 124, 125, 126, 127, 129, 153, 162, 163, 171, 172
magnetic resonance spectroscopy (MRS), vii, 7, 8, 11, 140, 156
magnetic source imaging (MSI), vii, 7, 8, 11, 12, 17, 22, 23, 29
magnetoencephalography (MEG), vii, 7, 8, 11, 12, 17, 48, 53, 54, 59, 60, 62, 63, 64, 106
major depression, 155, 156, 157
major depressive disorder (MDD), 130, 139, 140, 143, 144, 153, 155, 156, 157
malignancy, 116, 118, 119
mammalian tissues, 80
management, ix, x, 59, 111, 117, 127, 130
mandible, 118
manipulation, 24, 162
MANOVA, 137
manufacturer, 71
manufacturing, 70, 71, 75, 78, 79
mapping, vii, 52, 62, 63, 130, 131, 154, 155, 157, 171, 172
marrow, 80, 81, 82, 89, 90
matrix, 6, 8, 18, 33, 34, 45, 73, 80, 82, 83, 85, 86, 88, 163
maturation, 82
maximum intensity projection (MIP), 35, 36
measurement, 4, 22, 34, 35, 46, 154
measures, 45, 65, 94, 95, 102, 104
mechanical properties, 73, 85
media, 71, 85
medical care, 3, 80
medication, 100, 142
medicine, vii, x, 5, 6, 84, 87, 88, 130, 139, 152, 155
medulla, 9, 23, 121
medulla oblongata, 9
melancholic, 144
melanoma, 119
melt, 77, 78
melting, 77, 78, 79
membranes, 114
memory, 62, 99, 102, 106, 107, 108, 141, 146, 147, 151, 152, 162

memory performance, 106, 107
memory retrieval, 107
meningioma, 118, 119
mental disorder, 100
mental illness, 94, 96, 100, 101, 107
mesenchymal stem cells, 71, 81, 90
meta-analysis, vii, 7, 12, 25, 140, 155
metabolic changes, vii, x, 7, 12, 17, 130
metabolic dysfunction, 146
metabolic syndrome, 90
metabolism, 1, 132, 135, 137, 140, 146, 154, 156, 158
metabolites, 73
metals, 76, 78
metastasis, 116, 118
methylcellulose, 82
Mexico, 93, 100
mice, 89
microfabrication, 86
microscopy, vii
microstructure, 78
midbrain, 9, 16, 19, 21, 23
migraine, 21, 29, 123, 156
migration, 6
mild cognitive impairment, x, 130
mineralized, 83
Minnesota, 100, 124
minority, 115
mitochondria, 135, 154
mitochondrial, 130, 135, 140, 141, 142, 145, 146, 155, 156, 158
mitochondrial DNA, 130, 140, 156
model system, 6
modeling, 49, 50, 53, 60, 61, 62, 63, 85, 87, 97, 98
models, viii, 2, 3, 4, 5, 8, 10, 16, 17, 23, 47, 49, 50, 51, 53, 58, 60, 64, 68, 70, 71, 72, 75, 82, 84, 97, 98, 99, 101, 102, 112, 135, 136, 155
modified polymers, 86
mold, 79
molecules, 5, 74
Monte Carlo, 59
mood, 139, 140, 146, 156, 158
mood disorder, 140, 156, 158
morbidity, ix, 1, 69, 70
morphine, 8
morphogenesis, 6
morphology, 57, 64, 73, 77, 81, 87, 88, 107
morphometric, 104
mortality, 1, 3, 155
mosaic, 153
mothers, 156
motion, viii, 39, 41, 42, 43, 44, 45, 46
motor system, 19, 155

motor task, 152, 172
moulding, 87, 88
mouse model, 140
movement, 22, 61, 149, 152, 158, 163, 164
mtDNA, 130
mucosa, 80, 89
multidimensional, 9
multiple myeloma, 90
multiple sclerosis, 116, 118, 126
multipotent, 81
multivariate, 105, 136, 137, 138, 153
muscle cells, 81
muscle weakness, 117, 118
muscles, 118
musculoskeletal, 89
mutations, 156
myelin, 112, 114
myeloma, 81, 90
myopathy, 146

# N

National Institutes of Health (NIH), 94, 100
National Science Foundation, 79
neck, 4, 6, 75, 76, 80, 82, 86, 88, 89, 117, 120, 121, 122, 123, 142
neocortex, 54, 61
nerve(s), ix, 13, 16, 20, 26, 29, 73, 80, 81, 86, 89, 109, 111, 112, 113, 114, 115, 116, 117, 118, 119, 120, 121, 122, 123, 124, 125, 126, 127
nervous system, 8, 13, 22, 25
Netherlands, 163
network, 8, 19, 22, 28, 77, 85, 95, 97, 99, 105, 133, 170
networking, 146
neural networks, 171
neural systems, 107
neural tissue, 16
neuralgia, ix, 111, 112, 113, 114, 115, 116, 117, 121, 123, 124, 125, 126, 127
neuritis, 118, 127
neurobiological, x, 131, 152, 157
neurobiology, 26, 94, 147
neurodegenerative diseases, x, 131
neurodegenerative disorders, ix, 130
neuroimaging, ix, x, 47, 53, 67, 100, 101, 103, 128, 131, 136, 147, 150, 152, 153, 157, 159
neuroimaging techniques, ix, 47, 53, 136
neurological deficit, 112
neurological disease, 140
neurological disorder, 118
neuroma, 116
neuronal circuits, 48

neuronal density, 104
neuronal plasticity, 141
neuronal systems, 141
neurons, 9, 13, 23, 81, 94, 109, 112, 138, 145, 158
neuropathic pain, 10, 16, 20, 22, 27, 28
neuropathologic changes, 102
neuropathy, 26, 118, 119, 120, 127
neurophysiology, 62, 101
neuroprotection, 1, 2
neuroprotective, 1
neuropsychology, 101
neuroscience, 5, 106
neurotransmitter, 140
neutrophils, 1
New Jersey, 59
New Mexico, 93, 100
New York, 27, 46, 66, 155
nightmares, 146
nitrogen, 131
nociception, 25, 28
nociceptive, 9, 13, 28, 122
noise, viii, 31, 34, 39, 65, 163, 171
non-linear, 49, 55
non-linear dynamics, 55
nonverbal, 151
norepinephrine, 155
normal distribution, 136
Norway, 47
nucleic acid, 5
nucleus(i), 9, 16, 22, 23, 24, 109, 117, 118, 120, 121, 123, 151, 171
nutrients, 73, 74, 82

## O

observations, 13, 112, 126, 135, 136, 137
occipital cortex, 52, 57
occipital lobe, 54, 57, 67, 143, 144, 149
oculomotor nerve, 118
oil(s), 32, 33, 35, 39, 163, 171
omega-3, 141
operator, 1, 34, 132
ophthalmoplegia, 118, 119
opioid, 26
optic nerve, 118, 127
optic neuritis, 118, 127
optical imaging, vii
optimization, viii, 41, 42, 43, 45, 46, 86
oral cavity, 122
orbit, ix, 109, 118
orbitofrontal cortex, 54, 151
organ, 73, 80, 82, 84, 97, 98
organization, 6, 28, 138

orientation, 53
oxidants, 1
oxidation, 156
oxidative stress, 146
oxygen, 1, 74, 82, 131, 146, 154, 158, 163, 171, 172
oxygenation, 21, 172

## P

pain, vii, viii, ix, 7, 8, 9, 10, 11, 12, 13, 14, 16, 17, 18, 19, 20, 21, 22, 23, 24, 25, 26, 27, 28, 29, 30, 111, 112, 114, 115, 116, 117, 118, 119, 120, 121, 122, 123, 124, 125, 126, 127, 128, 141, 142, 153
pancreatic, 142
panic attack, 160
panic disorder, 160
parameter, viii, 41, 42, 43
parathyroid hormone, 156
parietal cortex, x, 9, 24, 29, 131, 148, 168
parietal lobe, 57, 67, 143, 144, 148, 149
Parkinson Disease (PD), ix, x, 65, 67, 130, 131, 140, 145
parotid gland, 120, 123
partial seizure, 67
particles, 75, 76, 77, 78, 82
passive, 105
pathogenesis, 103, 125, 140
pathogenic, 94
pathology, 20, 67, 112, 114, 118, 120, 122, 123, 133, 146, 158
pathophysiological mechanisms, 112
pathophysiology, 2, 28, 64, 66, 94, 102, 131, 155
pathways, 9, 19, 54, 56, 88
patient care, 2
patient management, x, 130
pediatric, 62
PEEK, 76
penalty, 32
perception(s), 25, 26, 151, 162
performance, 39, 66, 96, 98, 99, 102, 105, 106, 107, 137, 164
perfusion, 1, 132, 138, 146, 153, 154, 157, 158, 159, 160
peripheral blood, 81, 89
permeability, 49
permit, 25, 49, 74
personal, 71, 142, 151
personal computers, 71
personality, 142, 156
personality traits, 156
PET, vii, ix, 7, 8, 11, 17, 21, 23, 24, 25, 27, 28, 29, 48, 53, 58, 59, 61, 62, 68, 106, 107, 130, 131, 132, 137, 139, 142, 147, 150, 153, 154, 155, 159

PET scan, 107, 154
pH, 68, 82, 146
pharmaceutical, 86, 112
pharmacogenetics, 155
pharmacological, 2, 150
pharmacological treatment, 150
pharynx, 123
phenomenology, 102, 103
phenotype(s), 94, 140, 156
phosphorous, 156
photon(s), ix, 63, 130, 133, 134, 156, 157, 159, 160
photopolymerization, 75
physico-chemical properties, 90
physiology, 8, 64
physiopathology, iv
placebo, 158
placenta, 81
planning, 74, 75, 100, 151, 162, 172
plaque, 3, 4
plasticity, 26, 81, 141, 156
pleasure, 139
plexus, 122
point spread function, viii, 41, 42, 43
police, 158
polyester, 77
polymeric materials, 5
polymer(s), 74, 75, 76, 77, 78, 79, 85, 86, 87
polymorphism, 102
polypropylene, 76
pons, 112, 114, 124, 125
poor, 6, 17, 35, 113, 119, 142
population, 94, 95, 98, 114, 116, 118, 126, 132, 141, 147, 155, 158, 159
porosity, 70, 76, 77, 85
portability, 109
positron, ix, 27, 28, 29, 63, 97, 130, 131, 154, 159
positron emission tomography, ix, 27, 28, 29, 63, 97, 130, 159
posterior cortex, 165, 170
posttraumatic stress, 158, 159, 160
post-traumatic stress disorder (PTSD), x, 129, 130, 131, 146, 147, 148, 149, 150, 151, 152, 158, 159, 160
power, 56, 57, 65, 160
precipitation, 78, 140
prediction, 66
predictors, 127
preference, 39, 114
prefrontal cortex, 9, 16, 27, 29, 54, 107, 148, 151, 152, 160, 162
pressure, vii, 8, 14, 23, 30, 77, 79
prevention, 1, 3
primary brain tumors, 171

primate, 18, 87
principal component analysis (PCA), 130, 137, 138, 153
probability, 53, 95, 137
probe, 98
processing stages, 9
production, 74, 75, 79, 88, 140
progenitor cells, 80, 90
prognostic marker, 110
program, 85, 164
proliferation, 5, 71
promote, 73, 82
propagation, viii, 47, 53, 54, 55, 56, 57, 58, 59, 61, 64, 65
prophylaxis, 155
propylene, 159
protein immobilization, 86
protein(s), 5, 74, 76, 82, 86, 91, 135, 140
protocol(s), viii, 31, 33, 34, 42, 87, 107, 111, 153, 163
provocation, 147, 159
psychiatric disorders, ix, x, 95, 130, 131, 132, 133, 134, 135, 153, 158
psychiatric illness, 102
psychiatric patients, x, 130
psychiatrist(s), 101, 152
psychopathology, 101
psychosis, 107
psychosomatic, 103
psychotherapy, 160
pulses, 33
PVA, 76

## Q

qualitative differences, 24
quality assurance, 107
questionnaire, 124

## R

radiation, 161
radio, 114
radiopharmaceutical(s), x, 130, 131, 132, 149, 150, 154
rain, 8, 94, 107
range, 3, 44, 46, 57, 71, 76, 77, 78, 82, 118
rapid prototyping (RP), ix, 25, 69, 70, 71, 72, 74, 75, 76, 77, 79, 84, 85, 86, 90, 127
rat, vii, viii, 7, 8, 10, 12, 14, 17, 18, 21, 22, 23, 25, 26, 27, 28
ratings, 20

reaction time, 102, 164
reactive oxygen, 1
reactivity, 146
reading, 134
real time, 41
reality, 75
recall, 99, 147, 158
receiver operating characteristic (ROC), 130, 137, 155
receptors, 9, 13, 21, 22, 82
recognition, 151
recollection, 160
reconstruction, ix, 5, 51, 52, 61, 63, 65, 69, 71, 72, 74, 80, 83, 84, 85, 86, 87, 88, 89, 90, 125, 133
recovery, 109
recurrence, 110, 112, 114, 155
redistribution, 4
redox, 135
reduction, viii, 18, 31, 33, 36, 37, 38, 74, 151, 157, 165
refractory, 62, 112, 116, 126
regeneration, 86, 90
regenerative medicine, 6, 84, 87
region of interest (ROI), 34, 36, 52, 57, 130, 132, 137, 143, 144, 157
regional, x, 2, 4, 14, 20, 21, 27, 28, 29, 130, 131, 132, 134, 136, 137, 145, 147, 150, 152, 153, 154, 155, 156, 157, 158, 159
regional cerebral blood flow (rCBF), 4, 20, 22, 28, 130, 131, 133, 137, 138, 146, 147, 153, 154, 156, 157
regional cerebral blood volume (rCBV), 130
regional cerebral metabolic rates of glucose (rCMRGlu), 130, 132
regulation, 26, 146, 151, 152
rejection, 80
relationship(s), vii, ix, 5, 49, 80, 93, 94, 97, 98, 108, 115, 125, 126, 136, 137, 140, 153, 154, 157, 158
relatives, 94, 102, 108, 142, 156
relaxation, 32, 39, 152
relaxation times, 32, 39
relevance, 65, 137
reliability, 63, 101
remission, 142
remodelling, 73
repair, 80, 82, 83, 89
replication, 96
reprocessing, 149, 152, 158
reproduction, 6
resection, 48, 53, 55
resins, 75, 76
resolution, viii, 8, 17, 20, 25, 31, 32, 33, 35, 36, 37, 38, 39, 41, 42, 43, 44, 48, 55, 60, 61, 64, 71, 72,
76, 84, 87, 97, 112, 125, 131, 132, 133, 134, 136, 139, 150, 158, 171
resources, 3, 18, 70, 73, 81
respiratory, 140, 146
responsiveness, 13, 147
restructuring, 149, 151, 152
retention, 134, 135, 141, 142, 143, 144, 145, 146, 148, 149, 150, 153, 154
retina, 81
rheumatoid arthritis, 21, 27
rhythms, 55, 56, 65
rice, 64
right hemisphere, 95, 151, 167
risk, 2, 3, 4, 80, 94, 101, 102, 104, 108, 110, 114, 116, 118, 127, 146, 154, 155, 170
risk factors, 104, 155
robustness, 42
rodents, 88
Rome, 129
room temperature, 77, 78, 79

# S

safety, 2, 110, 112, 155
saline, 34
salt, 88
sample, 51, 100, 128, 132, 136, 137, 150, 155
sampling, 53, 61, 62, 131
saturation, 34, 163
savings, x, 130
scatter, 131, 134
scattering, 134
schizophrenia, ix, x, 62, 93, 94, 95, 96, 97, 98, 99, 100, 101, 102, 103, 104, 105, 106, 107, 108, 130, 131, 153, 158
schizophrenic patients, 62, 105
school, 101
science, 73, 85
sclera, 118
sclerosis, 61, 67, 115, 116, 118, 126
scores, 95, 142, 158
search, 11, 42, 94
searches, vii, 7, 11
seed, 168
seeding, 70, 71, 79, 82
segmentation, 55, 66, 71, 72
segregation, 138
seizure(s), 48, 54, 55, 56, 57, 58, 64, 65, 66, 67, 68, 140
selecting, 32, 43, 134
selective serotonin reuptake inhibitor, 140, 152, 160
self-assembly, 6
self-renewal, 81

semantic, 107
sensation, 9, 13, 22, 106, 112, 117, 118
sensitivity, ix, x, 2, 6, 13, 22, 23, 38, 39, 95, 109, 113, 114, 130, 133, 134, 137, 161, 162, 164, 168, 171, 172
sensitization, 8, 13, 16, 19, 21, 22, 26, 27, 29, 140
sensory modalities, 151
separation, 98, 105
series, 33, 56, 66, 101, 114, 115, 118, 124, 126, 163, 164
serotonin, 108, 130, 140, 152, 155, 160
serotonin transporter (SERT), 130
serum, 109, 156
severity, 139, 142
sex, 136
sexual abuse, 147, 150, 159
sexual violence, 149
shape, 2, 49, 52, 70, 74, 75, 79, 90, 132
shaping, 24
sharing, 153
side effects, 4
signal transduction, 6
signaling, 2, 82, 140
signaling pathway, 2
signals, 1, 5, 22, 24, 33, 48, 65, 66, 96, 122, 133, 152, 163
signal-to-noise ratio (SNR), viii, 31, 32, 33, 34, 35, 36, 37, 38, 39, 171
signs, vii, 7, 12, 17, 117, 118, 124, 127
similarity, 57
simulation, viii, 33, 35, 41, 42, 43, 44, 59
Singapore, 89
sintering, 74, 87
sinus(es), ix, 109, 111, 118, 122
sinusitis, 123
sites, 16, 21, 22, 49, 100, 101, 125
skeletal muscle, 156
skin, 13, 16, 21, 22, 23, 30, 77, 80, 81, 83, 151
small animal imaging (SAI), vii, viii, 7, 8, 10, 11, 25
smooth muscle, 109
smoothing, 72, 164
society, 147, 158
software, 72, 132, 133, 136, 164
solidification, 75, 78
solvents, 77, 79
somatic symptoms, 140, 153, 160
somatosensory, 9, 16, 20, 21, 22, 23, 29, 141
sounds, 80, 148
spatial information, 32, 98
spatial location, 61
spatiotemporal, 54, 55, 62
specialized cells, 80
species, 1, 90

specificity, ix, 95, 102, 113, 114, 125, 130, 137
SPECT, vii, ix, x, 7, 8, 11, 48, 53, 59, 63, 129, 130, 131, 132, 133, 134, 137, 139, 141, 142, 143, 144, 146, 147, 148, 149, 150, 152, 153, 154, 156, 157, 158, 159, 160
spectroscopy, 32, 48, 68, 140, 156
spectrum, 10, 65, 150, 155, 160
speech, 84
speed, viii, 41, 42, 62
spin, 4, 125
spin labeling, 4
spinal cord, 9, 13
spine, 123, 145
squamous cell carcinoma, 119
stability, 71, 82, 84
stages, 9, 38, 146
standard deviation, 34, 43, 45, 141
standardization, 133, 136, 154
starch, 76, 87
statistical analysis, 52, 97, 153
statistical inference, 137
statistical parametric mapping (SPM), viii, 47, 58, 99, 130, 133, 136, 138, 143, 144, 145, 167
statistics, 52, 96, 103, 132, 136, 137, 164
status epilepticus, 58, 62
stem cell(s), 71, 80, 81, 82, 89, 90
stenosis, 3, 4
stereolithographie (SLA), 74, 75, 76
stimulus, 12, 13, 18, 20, 22, 25, 150, 163
storage, 151
strategies, vii, ix, 1, 5, 44, 69, 70, 71
stratification, 3
strength, 32, 38, 73, 76, 79, 168, 170, 171
stress, x, 2, 130, 140, 141, 146, 147, 151, 155, 156, 158, 159, 160, 171
stressful life events, 140, 155
stressors, 147
striatum, 19, 28, 152, 170
stroke, 1, 2, 3, 4, 110, 146
structural equation modeling, 98, 106
subarachnoid hemorrhage, 127
subgroups, 3, 4, 144
substance abuse, 150
substitutes, 73, 79, 80, 81, 84
substrates, x, 6, 63, 130, 135, 152
subtraction, 20, 133
suffering, 146, 150, 159
sulfate, 74
superior temporal gyrus, 54, 95
supply, 4, 82
suppression, 32, 35, 36, 39

surgery, ix, 1, 2, 3, 4, 6, 62, 63, 64, 67, 69, 70, 71, 75, 76, 80, 84, 86, 87, 88, 89, 91, 109, 110, 112, 114, 117, 125
surgical intervention, 124, 125
surgical resection, 48, 53
survival, 151, 160
survivors, 148, 158
susceptibility, 102, 141, 142, 171
susceptibility genes, 102
Sweden, 129, 154, 161, 162
swelling, 85
Switzerland, 1, 5, 47
symbols, 151
symmetry, 72
symptom(s), 3, 94, 101, 112, 113, 118, 120, 121, 124, 139, 140, 141, 147, 150, 152, 153, 155, 158, 159, 160
synaptic plasticity, 156
synaptophysin, 158
syndrome, vii, viii, 7, 8, 14, 21, 25, 27, 29, 81, 90, 102, 104, 107, 117, 118, 120, 121, 123, 127, 153, 156, 158
synthesis, 99
systems, vii, 6, 51, 72, 74, 85, 87, 94, 101, 106, 107, 131, 133, 138, 139, 140, 141, 151

## T

targets, 99
taxonomy, 12
$T_c$, 20, 156, 157, 158
technetium, 154, 157
technology, 2, 71, 79, 88, 135
teeth, 71
temperature, 20, 75, 77, 78, 79, 88
template matching, 46
temporal, 17, 20, 25, 39, 48, 52, 53, 54, 55, 56, 57, 58, 60, 61, 62, 63, 64, 65, 66, 67, 80, 87, 94, 95, 97, 98, 99, 118, 120, 138, 141, 143, 145, 147, 148, 149, 153, 171, 172
temporal arteritis, 118
temporal lobe, 53, 55, 56, 57, 61, 62, 63, 64, 65, 66, 67, 95, 99, 141, 145, 148, 149, 153, 172
temporal lobe epilepsy, 53, 55, 61, 62, 63, 64, 65, 66, 67
temporo-parietal cortex, x, 131
tendon, 71, 80
territory, 4, 123
test statistic, 52
thalamus, x, 9, 16, 17, 21, 22, 54, 99, 149, 161, 165, 166, 170
theory, 60, 81, 94, 107, 112, 155
therapeutic, vii, 2, 94

therapeutic interventions, vii
therapy, 2, 90, 117, 145, 149, 152, 157
thermal degradation, 77, 79
thermoplastic, 76
thinking, 172
threats, 146
threshold(s), x, 12, 13, 34, 136, 140, 161, 164, 165, 166, 167, 170
thyroid gland, 123
tic douloureux, 127
time, viii, x, 2, 3, 6, 13, 17, 18, 22, 31, 32, 33, 35, 37, 38, 39, 41, 42, 43, 45, 46, 55, 56, 60, 66, 76, 77, 78, 80, 81, 82, 84, 89, 101, 102, 125, 130, 132, 138, 150, 153, 163, 164, 165, 170, 171
time resolution, 150
timing, 1, 43, 164
tinnitus, 141, 142, 144, 153, 157
tissue, vii, ix, 1, 5, 6, 8, 16, 19, 32, 69, 70, 71, 72, 73, 74, 75, 79, 80, 81, 82, 83, 84, 85, 86, 87, 88, 89, 90, 91, 110, 116, 121, 124, 135, 140
tissue engineering (TE), ix, 5, 33, 34, 69, 70, 71, 73, 74, 75, 76, 79, 80, 81, 82, 83, 84, 90, 125
titanium, 70
Tokyo, 125
tonic, 24, 30
totipotent, 81
trachea, 80, 123
tracking, 65, 98
trade-off, 38
trading, viii, 31
training, 18, 87, 163, 172
traits, 156
trajectory, 42
transcranial magnetic stimulation, 157
transduction, 6
transformation, 50, 125
transgenic, 10
transient ischemic attack, 110
transition, 73, 75, 78, 79
transition temperature, 75, 79
translation, 2, 50
transmission, 71, 112
transplantation, 88
transport, 73
transverse section, 34
trauma, 20, 146, 147, 148, 150, 151, 152, 158, 160
traumatic events, 150, 152
traumatic experiences, 146, 151
trend, 17, 142
tricyclic antidepressant, 142
trigeminal, ix, 16, 22, 29, 109, 111, 112, 113, 114, 115, 116, 117, 118, 119, 120, 121, 122, 123, 124, 125, 126, 127, 128

# Index

trigeminal nerve, ix, 109, 112, 113, 115, 116, 118, 120, 121, 122, 123, 124, 125, 126, 127
trigeminal neuralgia, ix, 111, 112, 113, 114, 115, 116, 117, 121, 124, 125, 126
trigeminal system, 29
trigemino-cardiac reflex (TCR), ix, 109, 110
tuberous sclerosis, 61
tumor(s), 66, 68, 80, 115, 118, 119, 122, 126, 127, 171
turnover, 140
tympanic membrane, 80, 85, 122

## U

UK, 111, 164
ultrasound, vii
umbilical cord, 81
uncertainty, 141
undifferentiated cells, 81
uniform, 34, 49
United States, 108
univariate, 135, 136, 138
urban population, 159

## V

vagus nerve, 109, 122
validation, 64, 113
validity, vii, 7, 11, 104
values, viii, 31, 32, 33, 38, 45, 71, 72, 110, 134, 136, 165
variability, 136, 137, 151, 157, 160
variable(s), 2, 57, 99, 132, 135, 136, 137, 150
variance, 34, 129, 132, 135, 164
variation, viii, 41, 43, 77, 101, 102, 108, 136, 137
vascular dementia, x, 130
vascular diseases, 32
vasculature, 38, 73
vasculogenesis, 90
vein, 81
verbal fluency, 105
vertebral artery, 128
vessels, 36, 38, 80, 81, 112
vestibular schwannoma, 126

vestibulocochlear nerve, 120
veterans, 150, 159
victims, 80, 146
Vietnam, 159
violence, 149, 158, 159
viral infection, 120
virus(es), 118, 127
viscera, 109
vision, 118, 162
visual acuity, 118
visual perception, 162
visual processing, 97
visual stimuli, 151
visualization, 133
vitamin D, 141
volume of interest (VOI), 130, 132, 136, 137, 160
vulnerability, 3, 102, 140

## W

Washington, 158
wavelet analysis, 55
weakness, 117, 118
welfare, 25
white matter, 50, 98, 118, 132
windows, 117
withdrawal, 146
women, 150, 156, 159, 162
workers, 19, 23
working memory, 99, 102, 106, 107, 108, 141, 162
workstation, 34, 72

## X

X-ray, 38

## Y

yield, viii, ix, 41, 53, 59, 111, 112, 118, 123, 127, 128, 137
young adults, 159